Springer Series in Microbiology

Editor: Mortimer P. Starr

Basidium and Basidiocarp
Evolution, Cytology, Function, and Development

Edited by
Kenneth Wells
Ellinor K. Wells

With Contributions by
G. W. Gooday, H. E. Gruen, T. Ishikawa, B. C. Lu,
D. J. McLaughlin, F. Oberwinkler, C. Thielke, I. Uno

With 117 Illustrations

Springer-Verlag
New York   Heidelberg   Berlin

Kenneth Wells
Department of Botany
University of California
Davis, California 95616/USA

Ellinor K. Wells
803 Pine Lane
Davis, California 95616/USA

Series Editor:
Mortimer P. Starr
Department of Bacteriology
University of California
Davis, California 95616/USA

Production: Kate Ormston

Library of Congress Cataloging in Publication Data
Main entry under title:
Basidium and basidiocarp.
  (Springer series in microbiology)
  Bibliography: p.
  Includes index.
  1. Basidiomycetes. I. Wells, Kenneth, 1927–
II. Wells, E. K. III. Gooday, G. W., 1942–
IV. Series.
QK626.B37   589.2′2   81-14580   AACR2

© 1982 by Springer-Verlag New York Inc.
All rights reserved. No part of this book may be translated or reproduced in any form without written permission from Springer-Verlag, 175 Fifth Avenue, New York, New York 10010, U.S.A.
The use of general descriptive names, trade names, trademarks, etc., in this publication, even if the former are not especially identified, is not to be taken as a sign that such names, as understood by the Trade Marks and Merchandise Marks Act, may accordingly be used freely by anyone.

Printed in the United States of America

9 8 7 6 5 4 3 2 1

ISBN 0-387-90631-2 Springer-Verlag New York Heidelberg Berlin
ISBN 3-540-90631-2 Springer-Verlag Berlin Heidelberg New York

Preface

The intent of this publication is to bring together reviews and discussions from several disciplines, all treating the basidium and basidiocarp of the Basidiomycotina (= basidiomycetes), a subdivision of the true or higher fungi. Because the workers who study the species of this group employ such a variety of techniques and publish in such diverse journals, we believe that bringing together these efforts in one publication will facilitate a synopsis of recent studies of several divergent disciplines. Correlation of such information may not only aid in the reevaluation of broad taxonomic and biological concepts but also provide a key to the specialists in the rethinking of the data available within the confines of the more restricted disciplines. We have attempted to cover the major areas of studies of species of the Basidiomycotina within the past decade or so with the exception of genetics and compatibility, which have recently been reviewed in several other works.

A problem we have not been able to solve satisfactorily is the one of vocabulary. Each discipline tends to develop its own language as it becomes increasingly specialized, with time becoming unintelligible to the majority. We have tried to alleviate this problem of terms but can not claim to have been completely successful.

We are indebted to a great many people, but especially to the contributors. They have been most patient and cooperative throughout. We also wish to thank Mortimer Starr, as well as Kate Ormston and Mark Licker of Springer-Verlag, for their advice and help.

We are most grateful for permission from Gustav Fisher Verlag to reproduce Figures 1.5-2, 1.9-3, 1.10-4, and 1.10-6, from the Deutsche Botanische Gesellschaft to reproduce Figures 1.16-1b and 1.16-3b, from the Deutsche Gesellschaft für Mykologie to reproduce Figure 1.10-5, and from Verlag Ferdinand Berger & Söhne Gesellschaft to reproduce Figures 1.16-2a, 1.16-2b, and 1.16-4b. Figures 2.13 and 2.15 are reproduced with the consent of the American Journal of Botany. Figures 4.2, 4.3, and 4.6 are reproduced with the permission of the Genetics Society of Canada, Figures 4.3, 4.8, and 4.14 with the permission of the Genetics Society of America, and Figures 4.4, 4.10, 4.12, 4.15, 5.1, and portions of Table 5.1 with the permission of Springer-Verlag Heidelberg. The Microbiology Research Foundation has authorized the reproduction of portions of Tables 5.5

and 5.3; and the American Society for Microbiology has granted permission to reproduce portions of Tables 5.2, 5.3, and 5.4. Finally Elsevier/North-Holland Biomedical Press has granted permission to reproduce Figures 5.2 and 5.4, parts of Figure 5.3 and portions of Table 5.5.

Davis, California
February, 1982

Kenneth Wells
Ellinor K. Wells

Contents

Introduction 1

K. WELLS

References 6

Chapter 1
The Significance of the Morphology of the Basidium in the Phylogeny of Basidiomycetes 9

F. OBERWINKLER

Introduction 9
Heterobasidiomycetes 10
Uredinales 12
Septobasidiales 12
Ustilaginales s. str. 15
Auriculariales 15
Tremellales 18
Holobasidiate Taxa 21
Basidiomycete Yeasts and Yeast-like Taxa 24
Homobasidiomycetes 28
References 34

Chapter 2
Ultrastructure and Cytochemistry of Basidial and Basidiospore Development 37

D. J. MCLAUGHLIN

Introduction 37
Basidial Initiation 38
Storage Product Accumulation 40
Wall Development in the Basidium 42
Basidial Septation 46
Sterigmal Initiation 48

Basidiospore Formation 49
Cytoplasmic Differentiation at the Septal Pore 53
Vacuolation of Maturing Basidia 54
Endomembrane System 55
Other Organelles 60
Experimental Control of Late Basidial Development 61
Materials and Methods 64
Summary and Conclusions 67
References 69

Chapter 3
Meiotic Divisions in the Basidium 75

C. THIELKE

Introduction 75
Methods 76
The Sequence of Meiosis 76
Discussion 85
References 90

Chapter 4
Replication of Deoxyribonucleic Acid and Crossing Over in Coprinus 93

B. C. LU

Introduction 93
An Overview of the System 93
Premeiotic DNA Replication 97
Genetic Recombination 99
Recombination, a Coordinated Program of the Meiotic Cell
 Cycle 108
Concluding Remarks 111
References 111

Chapter 5
Biochemical and Genetic Studies on the Initial Events of Fruitbody Formation 113

I. UNO AND T. ISHIKAWA

Introduction 113
Detection of Fruitbody-Inducing Substance 113
Regulation of Cyclic AMP Level and Fruiting 115
Existence of Cyclic AMP-Dependent Protein Kinases 117
Regulation of Glycogen Metabolism by Cyclic AMP 118
Possible Role of Cyclic AMP in Fruiting 120
References 122

Contents

Chapter 6
Control of Stipe Elongation by the Pileus and Mycelium in Fruitbodies of *Flammulina velutipes* and Other Agaricales **125**

H. E. GRUEN

Introduction   125
Role of the Pileus in Stipe Elongation of Agaricales   126
Relationship Between Stipe Elongation and the Mycelium   134
Plant Growth Regulators and Nucleotides in Relation to
   Fruitbody Growth   135
Recent Studies on *Flammulina velutipes*   137
Discussion   149
References   152

Chapter 7
Metabolic Control of Fruitbody Morphogenesis in *Coprinus cinereus* .   **157**

G. W. GOODAY

Introduction   157
Materials and Methods   157
Results   158
Discussion   168
Concluding Remarks   172
References   172

Author Index **175**

Subject Index **179**

Contributors

GRAHAM W. GOODAY — Department of Microbiology, University of Aberdeen, Aberdeen AB9 1AS, Scotland, United Kingdom

HANS E. GRUEN — Department of Biology, University of Saskatchewan, Saskatoon, Saskatchewan S7N 0W0, Canada

TATSUO ISHIKAWA — Institute of Applied Microbiology, University of Tokyo, Bunkyo-Ku, Tokyo, Japan

BENJAMIN C. LU — Department of Botany and Genetics, University of Guelph, Guelph, Ontario N1G 2W1, Canada

DAVID J. MCLAUGHLIN — Department of Botany, University of Minnesota, St. Paul, Minnesota 55108, USA

FRANZ OBERWINKLER — Chair for Special Botany and Botanical Garden, University of Tübingen, Auf der Morgenstelle 1, D-7400 Tübingen, Federal Republic of Germany

CHARLOTTE THIELKE — Institute for Plant Physiology and Cell Biology, Free University of Berlin, Königin-Luise Strasse 12-16a, D-1000 Berlin 33, Federal Republic of Germany

ISAO UNO — Institute of Applied Microbiology, University of Tokyo, Bunkyo-Ku, Tokyo, Japan

KENNETH WELLS — Department of Botany, University of California, Davis, California 95616, USA

Introduction

KENNETH WELLS

In recent years, scientific disciplines have become increasingly specialized. Not only do specialists of a particular discipline tend to communicate only with those of the same discipline, but investigators are inclined to fragment their contacts still further along lines of investigative techniques. In mycology, no less than in other sciences, subdivisions have developed that often parallel subsections of other sciences or are generated by new investigative techniques. During the past two decades, electron microscopy has made substantial contributions to fungal cytology; however, electron microscopic studies often ignore previous, and even current, studies by light microscopy. The studies of the influence of pileus diffusate on stipe elongation in several species of the Agaricales were influenced by the voluminous papers dealing with hormonal control of plant coleoptile curvature and/or elongation, especially those utilizing *Avena* coleoptiles. Often the isolation of investigators is further fostered by the policies of journals and societies. While the geometric increases of knowledge in many disciplines quite naturally dictate ever-increasing specialization, there is a need, as Andrews (1980) argues, for cross-discipline communication. There are significant advantages of peeking into the backyards of others; possibly even a visit over the fence would be instructive for many. It is the purpose of this volume to bring together reviews of several areas of recent studies of the basidiomycetes, especially those of the basidium and of the basidiocarp. The theme of this presentation is on the objects being studied by several techniques of research.

Early investigations of the Basidiomycotina were mainly taxonomic in nature (Ainsworth, 1976). Increases in the number of taxa and in the complexity of the taxonomic systems stimulated, and improvements of the magnifying systems made possible, a microscopic search for structures and features that would guide the taxonomists to more natural classifying systems.

Léveillé's (1837) clear descriptions and illustrations of the basidia and basidiospores of several species of the Homobasidiomycetes were a most significant advancement in both the taxonomy and cytology of the basidiomycetes. The classical studies of Tulasne (1853), Tulasne and Tulasne (1871), and Brefeld (1888) demonstrated the major basidial types in the Heterobasidiomycetes. These and similar studies formed the basis for Patouillard's (1887, 1900) thesis that the primary taxa of the basidiomycetes should be defined on the type of basidiospore germination and on basidial morphology. Recently, several authors (Talbot, 1965,

1968, 1970; Lowy, 1968, 1969; Ainsworth, 1973; McNabb, 1973; McNabb and Talbot, 1973) have proposed other major basidiomycete taxa emphasizing the taxonomic importance of internal basidial segments and teleutospores (i.e., thick-walled, resistant probasidia; see Oberwinkler, Chapter 1); however, I believe that these recent proposals disregard the fundamental biological differences between the heterobasidium and homobasidium, as they were defined by Patouillard (1900), and between the typical basidiocarps of the Heterobasidiomycetes and those of the Homobasidiomycetes. There is also a tendency, it seems to me, to base phylogenetic hypotheses on definitions rather than on the sum total of the stable characteristics of the various taxa. It seems likely, also, that the economic importance of the Uredinales, Tilletiales, and Ustilaginales has influenced their taxonomic treatments; the terminology applied to the basidia of these groups has not always, as Oberwinkler (Chapter 1) notes, properly emphasized the morphological and cytological similarities with the basidia of the saprobic Heterobasidiomycetes. Taking into account recent studies, Oberwinkler (Chapter 1) essentially proposes a return to Patouillard's (1900) taxonomic system with the primary emphasis on methods of basidiospore germination and basidial morphology. The data available from cytological, ultrastructural, and genetic investigations are most logically compatible with the phylogenetical hypotheses described and diagramed by Oberwinkler (Chapter 1). One great advantage of Patouillard's system is its flexibility, which permits obviously related organisms to be included within the same taxon (Martin, 1945). Other systems tend to rely more on definitions and/or on single, rigid characters (e.g., internal basidial segmentation). Heterobasidia, defined as deeply divided basidia or as basidia with internal segments, are significantly different biologically from the highly specialized and internally synchronized homobasidia.

Oberwinkler's proposals (Chapter 1) on basidiomycete phylogeny demonstrate the need for further refined studies of basidial and basidiospore ontogeny. While a great deal remains to be learned with critical light microscopic studies of many taxa, and many taxa remain to be discovered, the electron microscope has proven to be a boon to cytological investigations of these structures. McLaughlin (Chapter 2) reviews the non-nuclear events during basidial and basidiospore ontogeny. Thielke (Chapter 3) examines recent studies of meiosis in the basidium, especially those dealing with the spindle apparatus and associated structures.

One of the most significant reports on basidial and basidiospore ontogeny was that by Corner (1948). This study, and the earlier studies of de Bary (1866) focused attention on the basidium's role as a charged ampoule whose content is emptied into the basidiospores, in addition to its role as the site of meiosis and as a taxonomic indicator. The high resolving power of the electron microscope and more refined microchemical techniques have provided the means of extending the perceptive light microscopic observations of de Bary (1866), Corner (1948), and others. McLaughlin (Chapter 2) reviews the recent light and electron microscopic observations on basidial and basidiospore ontogeny, paying special attention to his studies of *Coprinus cinereus* (Schaeff. ex Fr.) S. F. Gray and *Auricularia fusco-succinea* (Mont.) Farl. He presents the cytochemical and ultrastructural basis for the presumed role of the basidium as the site of synthesis of storage nutrients (i.e., glycogen and/or lipids) of the basidiospores. As Oberwinkler (Chapter 1) points out, it is important that like stages of basidial ontogeny be properly equated in the saprobic and parasitic taxa if their real relationships are to be understood. McLaughlin's own studies, and those he discusses, help to interpret the thick-walled resistant teliospore (= teleutospore) of the Ustilaginales, Tilletiales, and Uredinales in relation to the thin-walled, more ephemeral probasidia of the saprobic Heterobasidiomycetes. Available studies of teliospore wall formation support the hypothesis that such structures have evolved separately in several taxa. Further, more detailed ultrastructural and cytochemical studies of teliospores at various stages of ontogeny will supply evidence on the ecological role of such structures. McLaughlin (Chapter 2) also reviews and discusses basidial and basidiospore wall development. A clearer understanding of these processes in a variety of taxa will add much to the reliability of theories pertaining to the phylogeny and origin of the Basidiomycotina.

While ultrastructural and cytochemical studies of the basidium have added much to the understanding of the basidium, much remains to be added. The critical ultrastructural studies, often of serial sections, accompanied by refined cytochemical tests and experimental treatments described by McLaughlin (Chapter 2) offer the means of gaining the information if applied to a variety of basidial types and to a sufficient number of the stages of development.

Some cytological features of the basidium recently uncovered in studies with the electron microscope were completely unexpected on the basis of past light microscopic studies. One such feature is the outer cap, first noted by Thielke (1972) and subsequently by several others, situated near the septal pore apparatus of the sub-basidial and subhymenial septa of a number of species of the Homobasidiomycetes. Another such feature is the septal pore apparatus recently reported by McLaughlin (1979) in the septa within the basidia of *Auricularia fuscosuccinea*. McLaughlin (Chapter 2) describes and illustrates these features and reviews the proposals that have been advanced to account for their functions.

That the basidium is the site of karyogamy and meiosis was established by the studies of Rosenvinge (1886), Wager (1893, 1894), Dangeard and Sapin-Trouffy (1893), Rosen (1893), Dangeard (1895), and others. The relatively small size of basidiomycete nuclei, and their usual failure to stain with the usual staining procedures seriously hampers cytological studies in this group. The light microscopic studies of Wakayama (1930, 1932), Olive (1942), Lu (1964), and others established the essential features of chromosome morphology during meiosis, especially during prophase I. Electron microscopic investigations, especially those correlated with light microscopic studies, have stimulated renewed interest in cytological events in the basidium. Thielke (Chapter 3) discusses her own studies, and those of others, especially as they pertain to the spindle apparatus and the nuclear envelope and to their behavior during meiosis.

While the fundamental features of meiosis in the basidium are as reported for other eukaryotic organisms, the bipolar, centriole-like bodies (= spindle pole bodies) detected in a number of basidiomycetes, with the exception of some taxa of parasitic Heterobasidiomycetes, differ structurally from the polar bodies of the spindle apparatus in the Ascomycotina and other major taxa of the fungi (Wells,

1977). As Thielke (Chapter 3) describes, the globular ends of the spindle pole body pass to the opposite ends of the nucleus during division. In interphase nuclei, the chromatin seems to be attached to the spindle pole body, a feature clearly illustrated by Harper (1905) in the Ascomycotina. This unusual feature and the apparent asynchronous disjunction of the chromatids during anaphase are unusual characteristics that also occur in the Ascomycotina.

Although the electron microscope has proven to be a most useful tool to the fungal cytologists, additional studies of nuclear divisions in the Basidiomycotina are needed, as noted by Thielke (Chapter 3), to resolve present discrepancies in the published works. Light microcopic observations of living meiotic and mitotic nuclei are needed, possibly utilizing vital staining. Serial sectioning and flat embedding techniques for electron microscopy are still most useful procedures that have not been applied to a sufficient number of studies; and, as Thielke (Chapter 3) emphasizes, interpretations of electron micrographs must take into account fixation damage.

Lu's (Chapter 4) studies of meiosis in the basidia of *Coprinus cinereus* have extended the information available on this process to the molecular level. Lu has been able to demonstrate that meiosis is inhibited in basidiocarps at 35°C in continuous light but occurs normally at 25°C in continuous light or at 35°C in a diurnal light/dark cycle. The 2-h sensitive period to light and temperature occurs prior to karyogamy. Measuring the incorporation of ^{32}P in DNA, Lu and his associates determined that the sensitive period was the premeiotic S-phase, which occurs several hours prior to karyogamy in the developing basidia. It is probably this sensitivity of the premeiotic replication to light that partially accounts for meiosis taking place at night and is perhaps associated with the synchrony of basidial development in *C. cinereus* and other deliquescing species of the genus *Coprinus*.

Extending his studies of high and low temperature to other phases of meiosis, Lu (Chapter 4) noted that high temperature during the late S phase, karyogamy, and pachytene increased the frequency of recombination. Low temperature, however, increased recombination only when applied to pachytene. Lu proposes that high temperatures at late S phase, karyogamy, and pachytene induce an increase in nicking of DNA, whereas cold temperature inhibits the DNA repair mechanism. Both temperature extremes increase recombination, but by different mechanisms. The experiments by Lu and his associates using ^{32}P as a label of DNA synthesis support this hypothesis. Lu proposes that recombination is not a haphazard process but that systems have evolved to promote crossing over. Since the primary function of the basidium is the production of the perfect spores bearing the meiotic nuclei, systems promoting crossing over are of elementary significance for the understanding of the biology and evolution of the Basidiomycotina.

Uno and Ishikawa (Chapter 5) describe a series of studies suggesting that adenosine 3′5′-cyclic monophosphate is an essential component in the initiation of basidiocarp formation in a form of *Coprinus macrorhizus* Rea. Using several mutant strains of this species, Uno and Ishikawa present evidence that basidiocarp formation is dependent on the accumulation of cyclic AMP in the mycelium. They also present clues and discuss the possible roles of cyclic AMP in basidiocarp induction. While basidiocarp initiation is no doubt a very complicated process, the studies by Uno and Ishikawa provide an insight into the genetic and biochemical processes in this fundamental event.

The basidia-bearing bodies of the Basidiomycotina vary from the minute pustules of the species of *Puccinia* (Uredinales) and the flat, nearly imperceptible, gelatinous films of many species of *Exidiopsis* (Tremellales) to the large, perennial, applanate-to-ungulate basidiocarps of the species of *Ganoderma* (Aphyllophorales). In the basidiocarps of many "typical mushrooms" (Agaricales), the differentiation of the essential features are essentially completed in the button or unexpanded stage. When the environmental conditions are favorable, the stipe and pileus develop rapidly due primarily to hyphal segment elongation. This rapid expansion of the differentiated basidiocarp would seem to serve as a mechanism to elevate the basidia-bearing surfaces above the substrate during those relatively brief periods when environmental conditions support basidiospore maturation, discharge, and germination. Gruen (Chapter 6) has brought together the results of his own extensive studies, and those of others, on the influence of the pileus and somatic mycelium on stipe elongation during the rapid expansion stage. Essentially, he accounts for the reasons why mushrooms mushroom.

Certainly, many mushroom collectors must have wondered why basidiocarps with a damaged pileus develop a distorted stipe. Perhaps even a few have noted that the curvature of the stipe is correlated with the region of the damage or missing pileus. In the scientific literature, Schmitz (1842) is credited with first noting the importance of the developing pileus to stipe elongation and that the zone of elongation is essentially restricted to the apical region of the stipe. Later, workers further refined the understanding of the relationships between the pileus and stipe. Borriss' (1934a, 1934b) studies were the first that could be described as experimental. He also first suggested that stipe elongation was controlled by a hormonal mechanism. Subsequent studies by several workers demonstrated the role of the lamellae and pilar context in stipe elongation when the treatments were applied at different stages of basidiocarp expansion. Gruen (Chapter 6) also reviews and discusses the evidence that the somatic mycelium and/or nutrients enhance stipe elongation and that the numbers of hyphal segments increase during this phase.

While the earlier studies were made mainly with the basidiocarps of *Agaricus bisporus* (Lange) Imbach and of species of *Coprinus*, the difficulties of working with the relatively large basidiocarps of *A. bisporus* and of obtaining relatively large numbers of essentially uniform basidiocarps prompted Gruen (Chapter 6) to develop the straight growth test for lamellar diffusate with the stipes of *Flammulina velutipes* (Curt. ex Fr.) Sing. By testing different media, using standardized culture conditions, and rigorously selecting the test stipes, variation of the effects of various treatments has been drastically reduced. Utilizing the refined techniques, Gruen has re-investigated the roles of lamellae and pilar context on stipe elongation. He has been able to clearly demonstrate that stipe elongation is dependent on lamellae during the phase of rapid elongation of the stipe. Gruen discusses his studies, some of which are described here for the first time, to determine the effects of lamellar diffusate at various stages of ontogeny and of nutrient on stipe elongation excised from basidiocarps of various ages. He also describes his efforts to distinguish the effects of nutrients and lamellar diffusates, to detect the movement of metabolites when applied to different regions of the decapitated stipes, to identify substances capable of inducing stipe elongation, and to enumerate the chemical and physical properties of the active agent.

Complementing the studies described by Gruen is the report by Gooday (Chapter 7), who describes his efforts to determine which hyphal constituents control stipe elongation by analyzing chemically the major components before and after the major period of elongation. By detecting those constituents that increase, remain essentially constant, or decrease, and by correlating these studies with the use of ^{14}C-labeled N-acetylglucosamine, Gooday provides evidence of the biochemical changes that occur during stipe elongation.

References

Ainsworth, G. C.: Introduction and keys to higher taxa. In: The Fungi, Vol. IVB. Ainsworth, G. C., Sparrow, F. K., Sussman, A. S. (eds.). New York: Academic Press 1973, pp. 1–7.
Ainsworth, G. C.: Introduction to the History of Mycology. London: Cambridge University Press 1976.
Andrews, J. H.: Plant disease as a biological phenomenon. Bioscience *30*, 647 (1980).
Borriss, H.: Beiträge zur Wachstums- und Entwicklungsphysiologie der Fruchtkörper von *Coprinus lagopus*. Planta (Berl.) *22*, 28–69 (1934a).
Borriss, H.: Über den Einfluss äusserer Faktoren auf Wachstum und Entwicklung der Fruchtkörper von *Coprinus lagopus*. Planta (Berl.) *22*, 644–684 (1934b).
Brefeld, O.: Untersuchungen aus dem Gesammtgebiete der Mykologie. 7. Basidiomyceten II. Protobasidiomyceten. Leipzig: Arthur Felix 1888.
Corner, E. J. H.: Studies in the basidium. I. The ampoule effect, with a note on nomenclature. New Phytol. *47*, 22–51 (1948).
Dangeard, P. A.: Mémoire sur la reproduction sexuelle des Basidiomycètes. Botaniste *4*, 119–181 (1895).
Dangeard, P. A., Sapin-Trouffy, P.: Une pseudo-fécondation chez les Uredinées. Compt. Rend. Hebd. Séances Acad. Sci. *16*, 267–269 (1893).
de Bary, A.: Morphologie und Physiologie der Pilze, Flechten und Myxomyceten. Leipzig: Wilhelm Engelmann 1866.
Harper, R. A.: Sexual reproduction and the organization of the nucleus in certain mildews. Carnegie Inst. Wash. Publ. *37*, 1–104 (1905).
Léveillé, J. H.: Recherches sur l'hymenium des champignons. Ann. Sci. Nat. Bot., Sér. 2, *8*, 321–338 (1837).
Lowy, B.: Taxonomic problems in the Heterobasidiomycetes. Taxon *17*, 118–127 (1968).
Lowy, B.: Septate holobasidia. Taxon *18*, 632–634 (1969).
Lu, B. C.: Chromosome cycles of the basidiomycete *Cyathus stercoreus* (Schw.) de Toni. Chromosoma (Berl.) *15*, 170–184 (1964).
Martin, G. W.: The classification of the Tremellales. Mycologia *37*, 527–542 (1945).
McLaughlin, D. J.: Ultrastructure of the hymenium of *Auricularia polytricha*. Mushroom Sci. *10*, 219–229 (1979).

McNabb, R. F. R.: Phragmobasidiomycetidae: Tremellales, Auriculariales, Septobasidiales. In: The Fungi, Vol. IVB. Ainsworth, G. C., Sparrow, F. K., Sussman, A. S. (eds.). New York: Academic Press 1973, pp. 303–316.

McNabb, R. F. R., Talbot, P. H. B.: Holobasidiomycetidae: Exobasidiales, Brachybasidiales, Dacrymycetales. In: The Fungi, Vol. IVB. Ainsworth, G. C., Sparrow, F. K., Sussman, A. S. (eds.). New York: Academic Press 1973, pp. 317–325.

Olive, L. S.: Nuclear phenomena involved at meiosis in *Coleosporium helianthi*. J. Elisha Mitchell Sci. Soc. *58*, 43–51 (1942).

Patouillard, N.: Les Hyménomycètes d'Europe. Paris: Paul Klincksieck 1887.

Patouillard, N.: Essai Taxonomique sur les Families et les Genres des Hyménomycètes. Lons-le-Saunier: Duclume 1900.

Rosen, F.: Beiträge zur Kenntniss der Pflanzenzellen. II. Studien über die Kerne und die Membranbildung bei Myxomyceten und Pilzen. Beitr. Biol. Pflanz. *7*, 237–266 (1893).

Rosenvinge, L. K.: Sur les noyaux des Hyménomycètes. Ann. Sci. Nat. Bot., Sér. 7, *3*, 75–93 (1886).

Schmitz, J.: Mykologische Beobachtungen, als Beiträge zur Lebens- und Entwicklungsgeschichte einiger Schwämme aus der Klasse der Gastromyceten und Hymenomyceten. Linnaea *16*, 141–215 (1842).

Talbot, P. H. B.: Studies of *"Pellicularia"* and associated genera of Hymenomycetes. Persoonia *3*, 371–406 (1965).

Talbot, P. H. B.: Fossilized pre-Patouillardian taxonomy. Taxon *17*, 620–628 (1968).

Talbot, P. H. B.: The controversy over septate holobasidia. Taxon *19*, 570–572 (1970).

Thielke, C.: Die Dolipore der Basidiomyceten. Arch. Mikrobiol. *82*, 31–37 (1972).

Tulasne, E. L.: Observations sur l'organisation des Trémellinées. Ann. Sci. Nat. Bot., Sér. 3, *19*, 193–231 (1853).

Tulasne, E. L., Tulasne, C.: New notes upon the tremellineous fungi and their analogues. J. Linn. Soc. Lond., Bot. *13*, 31–42 (1871).

Wager, H.: On nuclear divisions in the Hymenomycetes. Ann. Bot. (Lond.) *7*, 489–514 (1893).

Wager, H.: On the presence of centrospheres in fungi. Ann. Bot. (Lond.) *8*, 321–334 (1894).

Wakayama, K.: Contributions to the cytology of fungi. I. Chromosome number in Agaricaceae. Cytologia (Tokyo) *1*, 369–388 (1930).

Wakayama, K.: Contributions to the cytology of fungi. IV. Chromosome number in Autobasidiomycetes. Cytologia (Tokyo) *3*, 260–284 (1932).

Wells, K.: Meiotic and mitotic divisions in the Basidiomycotina. In: Mechanisms and Control of Cell Division. Rost, T. L., Gifford, E. M., Jr. (eds.). Stroudsburg, Pa.: Dowden, Hutchison & Ross 1977, pp. 337–374.

Chapter 1

The Significance of the Morphology of the Basidium in the Phylogeny of Basidiomycetes

FRANZ OBERWINKLER

Introduction

The basidium is the organ in which karyogamy and meiosis occur and on which the meiospores (i.e., basidiospores) are developed. The structure of the mature meiosporangium (i.e., basidium) seems to be evolutionarily conservative; therefore, morphological similarities of the meiosporangia in the major taxa are believed to indicate varying degrees of relationships between such taxa. Since fossils of basidiomycete basidiocarps are almost completely lacking or are difficult to interpret, comparative morphological studies seem to offer the only feasible means of understanding evolution within the Basidiomycotina. The two major taxa of the subdivision Basidiomycotina are the classes Heterobasidiomycetes and Homobasidiomycetes. Included in the Heterobasidiomycetes are those species producing basidiospores that are capable of forming secondary spores and/or yeast-like cells. Those taxa lacking the ability to form secondary spores and/or yeast-like cells are included in the Homobasidiomycetes. The orders that I presently recognize in these two classes are enumerated in the following list. The genera mentioned are discussed in this contribution.

Heterobasidiomycetes

    Uredinales: *Chrysomyxa, Goplana, Ochropsora, Puccinia*
    Septobasidiales: *Septobasidium, Uredinella*
    Ustilaginales: *Leucosporidium, Rhodosporidium, Tilletiaria*
    Graphiolales
    Auriculariales: *Eocronartium, Herpobasidium, Hoehnelomyces, Paraphelaria, Phleogena, Pilacrella, Platygloea, Stilbum*
    Tremellales: *Hyaloria, Myxarium, Patouillardina, Protodontia, Pseudohydnum, Stypella, Tremellodendron, Tremellodendropsis, Tremiscus*
    Cryptococcales: *Filobasidiella, Filobasidium*
    Tulasnellales: *Ceratobasidium, Oliveonia, Pseudotulasnella, Tulasnella, Uthatobasidium, Ypsilonidium*
    Dacrymycetales: *Calocera, Cerinomyces, Dacrymyces*
    "Digitatisporales": *Digitatispora*
    Sporobolomycetales: *Aessosporon, Itersonilia, Sporidiobolus*

Exobasidiales: *Brachybasidium, Dicellomyces, Exobasidiellum, Exobasidium*
Tilletiales: *Tilletia*
Cryptobasidiales: *Coniodictyum, Microstroma*

Homobasidiomycetes

Aphyllophorales: *Albatrellus, Aphelaria, Athelia, Botryobasidium, Clavaria, Cyphellostereum, Humidicutis, Hydnopolyporus, Hyphoderma, Hyphodontia, Merulius, Mucidula, Multiclavula, Omphalina, Panellus, Paullicorticium, Phanerochaete, Phlebia, Porodisculus, Repetobasidium, Sistotrema, Stereophyllum, Tricholoma, Xeromphalina*
Cantharellales
Protohymeniales: *Vuilleminia, Xenasma*
Polyporales: *Lentinus*
Russulales: *Russula*
Hymenochaetales
Boletales
Agaricales: *Agaricus, Coprinus*
Hymenogastrales
Thelephorales
Lycoperdales
Geastrales
Sclerodermatales
Melanogastrales
Gastrosporiales
Gautieriales
Tulostomatales
Nidulariales
Phallales

Heterobasidiomycetes

Because they produce secondary spores, the Uredinales represent an order of the Heterobasidiomycetes (Fig. 1.1), a taxon that also includes, in my opinion, the Dacrymycetales, the Exobasidiales and, probably, the Cryptobasidiales. The main basidial types of the Heterobasidiomycetes are well known (Fig. 1.1) and are considered to be representative of each natural group. It is striking that both transversely and longitudinally septate basidia, as well as holobasidial meiosporangia, are present within these fungi. Moreover, it is apparent that in several major taxa species have evolved with passive spore liberation (i.e., with gastroid basidia). In the Ustilaginales, Cryptobasidiales, and Tilletiales, all taxa form gastroid basidia

Fig. 1.1. Survey of the characteristic basidial types of the orders of the Heterobasidiomycetes. Those orders with some species with yeast stages and those with some, or all, species forming gastroid basidia are indicated.

Significance of the Morphology of the Basidium

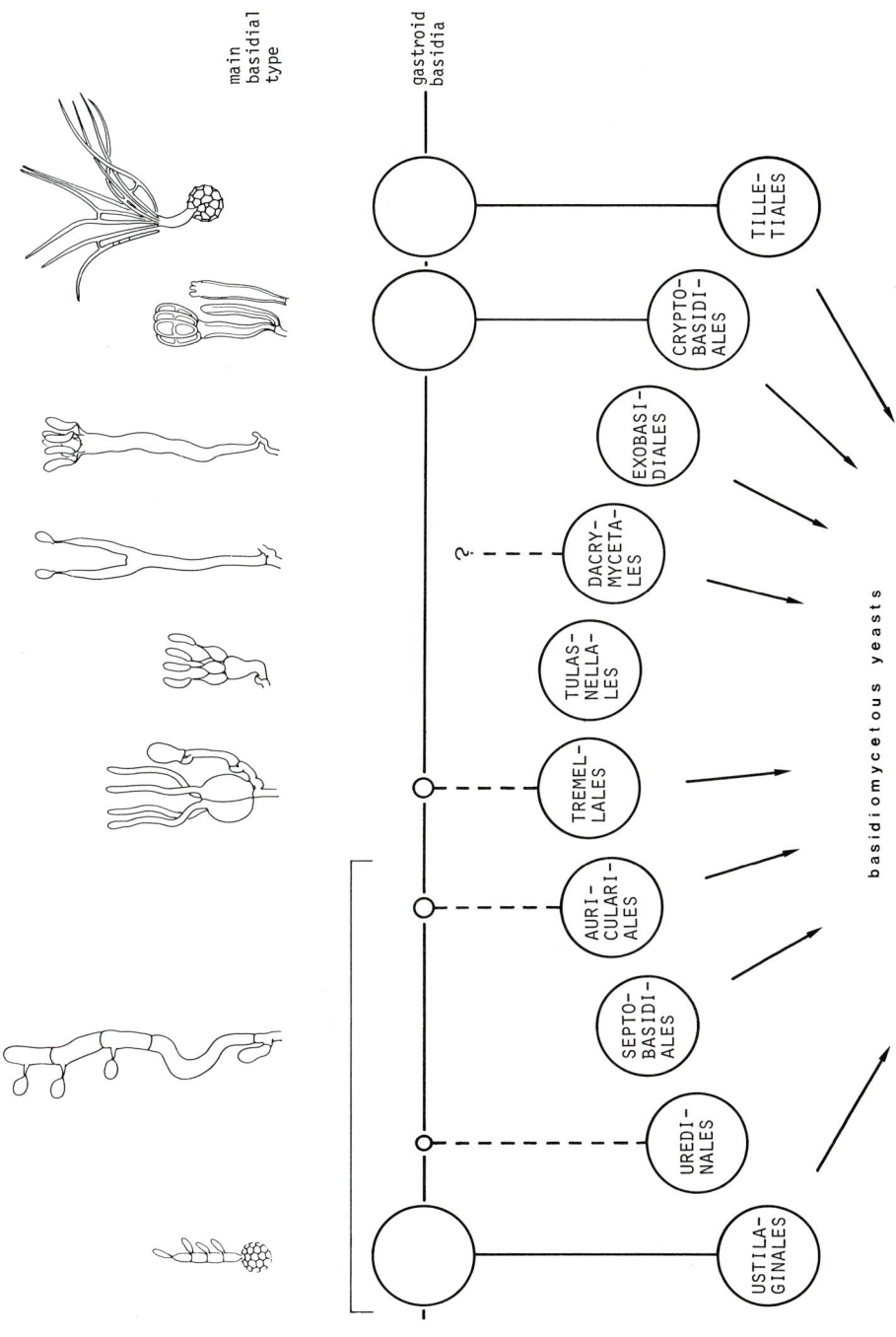

with passive spore liberation. Some gasteromycete-like representatives have evolved in the Uredinales, Auriculariales, and Tremellales. The Septobasidiales, Tulasnellales, and Exobasidiales seem to be only hymenomycetoid (i.e., with active spore liberation).

Uredinales

Although the rusts are highly specialized parasites on autotrophic, higher plants, some characteristics are fairly conservative: (i) their haploid phase is relatively long in comparison to other basidiomycetes; (ii) the rusts possess spermogonia and receptive hyphae, which are not found in other basidiomycetes; (iii) on the basis of our present knowledge, the perforations of the septa of the Uredinales are simpler and more ascomycete-like than the relatively complex dolipore septa of other basidiomycetes; and (iv) it seems likely that obligate parasitism and a high rate of asexual reproduction are primitive characters that do not commonly occur among the higher basidiomycetes.

For these reasons it is appropriate to place the rusts before other basidiomycetes in a linear system of classification. Within the Uredinales, basidia with a structure comparable to those of the Auriculariales can be found in the Pucciniastraceae, Coleosporiaceae, and Chrysomyxaceae (in the sense of Gäumann), which are generally considered as primitive rust taxa. In the species of *Goplana* and *Coleosporium* and in *Chrysomyxa abietis* (Wallr.) Ung. there are no probasidial swellings (Figs. 1.4-1, 1.4-2); the basidia are similar to those of *Eocronartium* of the Auriculariales and related taxa. Urediniologists describe such rust basidia as germinating internally; however, I think that basidial terminology should have a broader base to facilitate comparative discussions and studies in all auriculariaceous groups of the Heterobasidiomycetes. A sequence of host specificity within the Uredinales can be traced (Fig. 1.2) beginning on ferns, continuing through the gymnosperms, and terminating with the advanced rust taxa on the angiosperms. Very often this transition is accompanied by the appearance of morphologically distinct probasidia, which have evolved into thick-walled resting spores, teleutospores (e.g., *Puccinia caeomatiformis* Lagh., Fig. 1.4-3). On the other hand, hymenia or clusters of basidia that become enclosed during spore formation within the host tissue lose the capacity to form forcibly discharged basidiospores. Evidently this results in the evolution of gastroid basidia, as illustrated by *Ochrospora sorbi* (Oud.) Diet. (Fig. 1.4-4).

Septobasidiales

Uredinella coccidiophaga Couch (Couch, 1937) bridges the gap between the Uredinales and Septobasidiales. *U. coccidiophaga* possesses rust-like, thick-walled probasidia (Fig. 1.5-1) but is associated with scale insects and is not an obligate parasite on higher green plants. Within the Septobasidiales (Donk, 1964), meiosporangial variation resembles that known in the Uredinales. Thus, species with

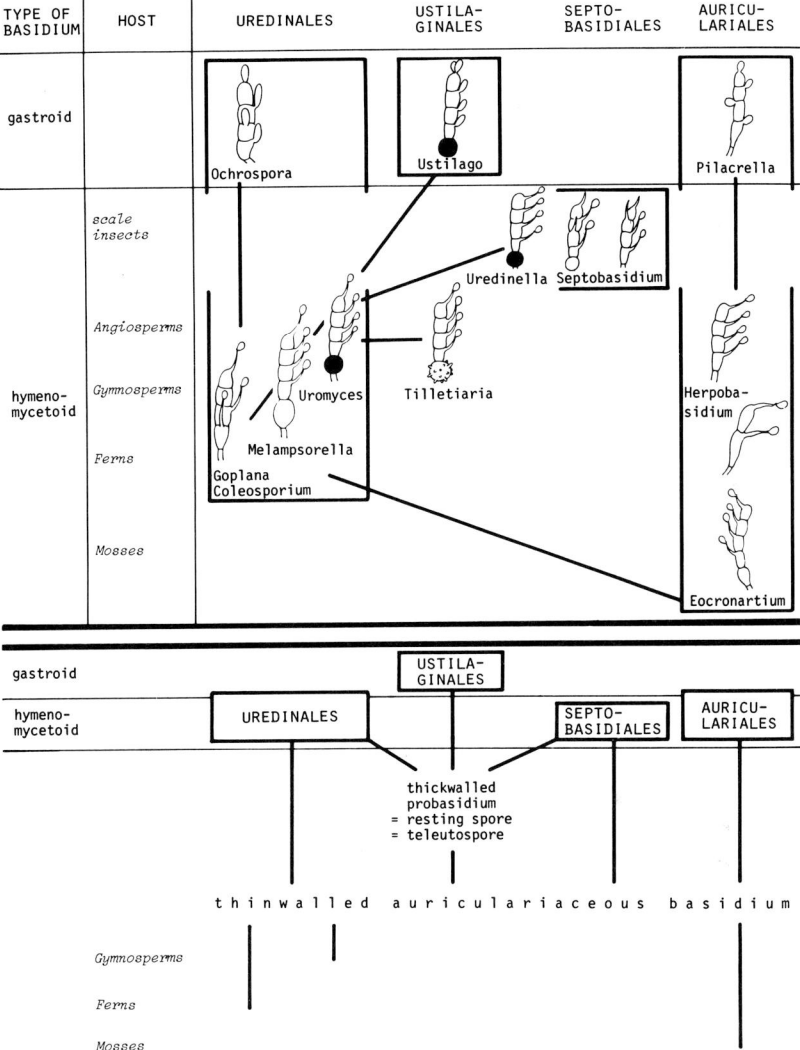

Fig. 1.2. A survey of the auriculariaceous basidium with its evolutionary tendencies. Indicated are (i) the phylogenetic lines of evolution of gastroid basidia within the Uredinales and Auriculariales, (ii) the evolution of the thick-walled probasidium (i.e., teleutospore) and its function as a resting structure, (iii) the correlation of the morphologically simple types of basidia with primitive land plants and their coevolution with the host plants, and (iv) the relative taxonomic position of the Septobasidiales and *Tilletiaria* based on basidial morphology.

and without morphologically distinct probasidia are included in the broadly conceived genus *Septobasidium*.

Possibly the auriculariaceous taxa originated on land plants and diverged in complexity following the evolution of the hosts (Fig.1.2). The thin-walled auriculariaceous basidium evidently has evolved to form those with a thick-walled

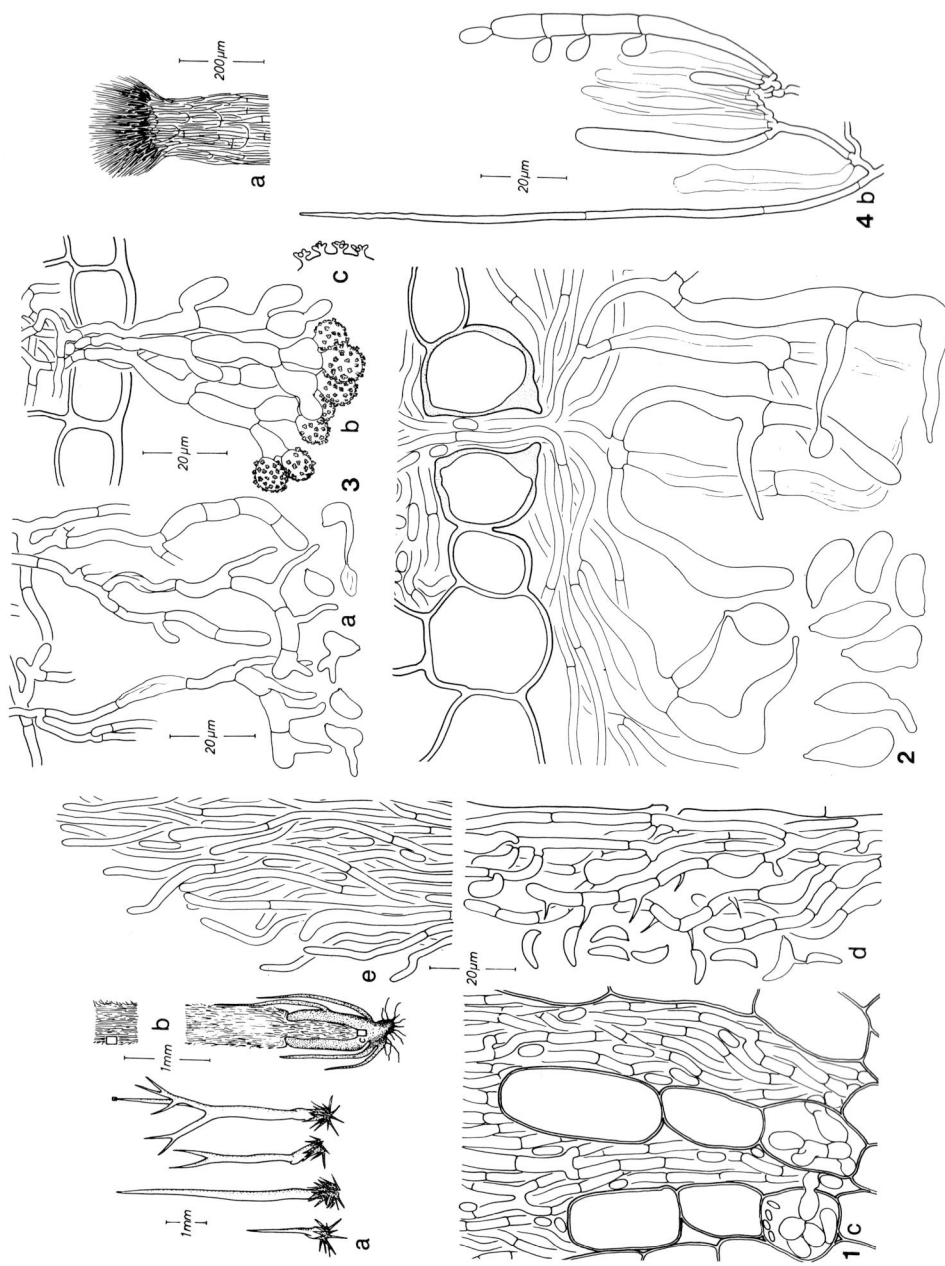

basidium, which is capable of functioning biologically as a resting spore. This differentiation in the function of the basidium (i.e., primarily as a spore-producing organ or primarily as a resting spore) may have separated the rusts, with forcibly discharged basidiospores, and the smuts, with a passive dispersal mechanism.

Ustilaginales s. str.

I know of no rust with gastroid metabasidia developing from well-developed teleutospores. The Ustilaginales s. str. (i.e., including only the Ustilaginaceae) may, however, represent fungi of this nature, at least those species without abnormal basidial and spore development. *Tilletiaria* (Bandoni and Johri, 1972) might be interpreted as an intermediate form between the Uredinales and the Ustilaginales, since in this genus typical smut spores are formed; but the basidiospores arise at the apices of curved sterigmata and are forcibly discharged.

Auriculariales

It has been postulated that the Ascomycotina and Basidiomycotina evolved simultaneously with the appearance of land plants. Present-day auriculariaceous parasites occur, *inter alia,* on mosses (Fig. 1.2); related taxa also occur on ferns and flowering plants (Fig. 1.2). The most highly evolved taxa possess stalked and capitate fruiting bodies enclosing gastroid basidia. The meiosporangia of all these organisms are similar in gross morphology; i.e., thin-walled, auriculariaceous basidia without morphologically distinct probasidia.

The clavarioid species of *Eocronartium* (Fig. 1.3-1) are parasitic on gametophytes of mosses, whereas in the tropics moss sporophytes are parasitized by species of *Jola* (Gäumann, 1922). *Herpobasidium filicinum* (Rostr.) Lind (Fig. 1.3-2; see Jackson, 1935), which is parasitic on green fern leaves, differs from these species by the formation of two-celled basidia. This does not seem to be a very important deviation since the closely related *Herpobasidium deformans* Gould (Fig. 1.3-3) (Gould, 1945) on Caprifoliaceae and Cornaceae normally develops

◀ **Fig 1.3.** Basidia of various taxa of the Auriculariales. **1.3-1** *Eocronartium muscicola* (Pers. ex Fr.) Fitzp. **a** Fruitbodies. **b** Section of fruitbody. **c** Interaction of host and parasite. **d** Portion of the hymenium with basidia and basidiospores. **e** Apical region of the basidiocarp. **1.3-2** *Herpobasidium filicinum.* A section through a portion of the leaf of the host [*Dryopteris filix-mas* (L.) Schott] showing the mycelium of the parasite, the two-celled basidia, and basidiospores. **1.3-3** *Herpobasidium deformans.* **a** Squashed fruitbody. **b** Conidial state. **c** Portion of a conidium wall. **1.3-4** *Pilacrella solani.* Section through the apical part of the fruitbody with basidia, basidiospores, and sterile hypha.

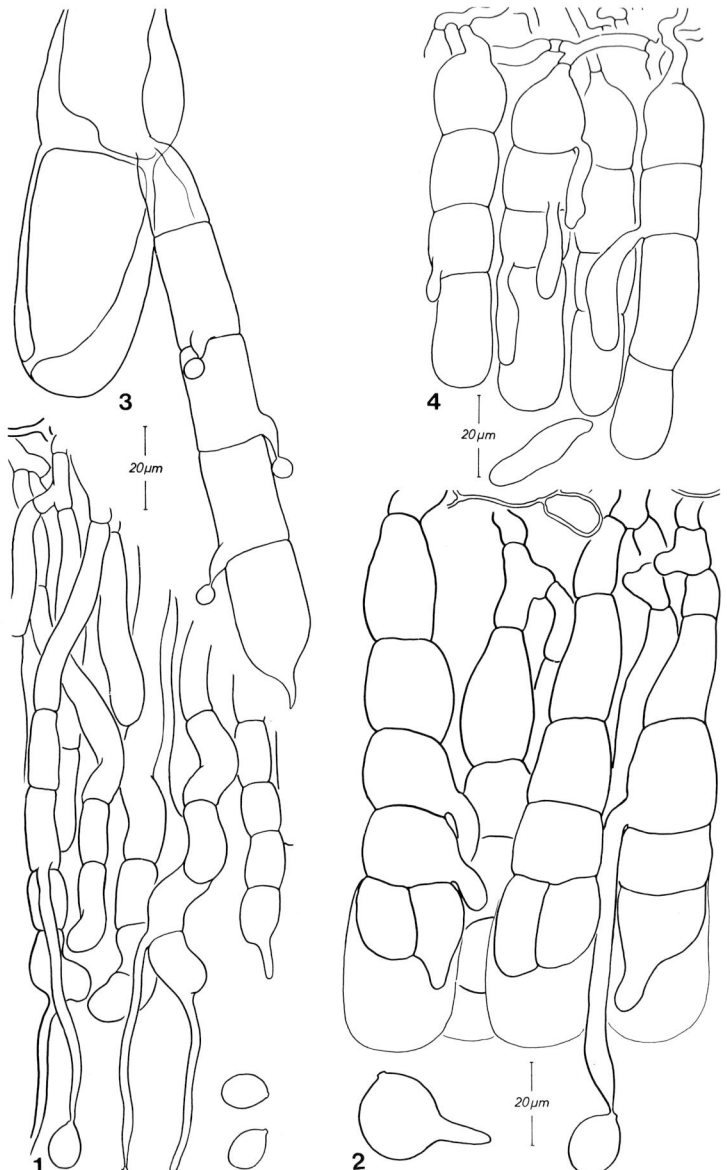

Fig. 1.4. Basidia of the Uredinales. **1.4-1** *Goplana micheliae* Racib. Basidia in various stages of development and basidiospores. **1.4-2** *Coleosporium sonchi* (Schum.) Lév. A portion of the basidial layer with basidiospores. **1.4-3** *Puccinia caeomatiformis* (on *Baccharis floribunda* H.B.K.). Germinating teliospore (i.e., thick-walled probasidium) with metabasidium. **1.4-4** *Ochropsora sorbi*. Gastroid basidia and basidiospores.

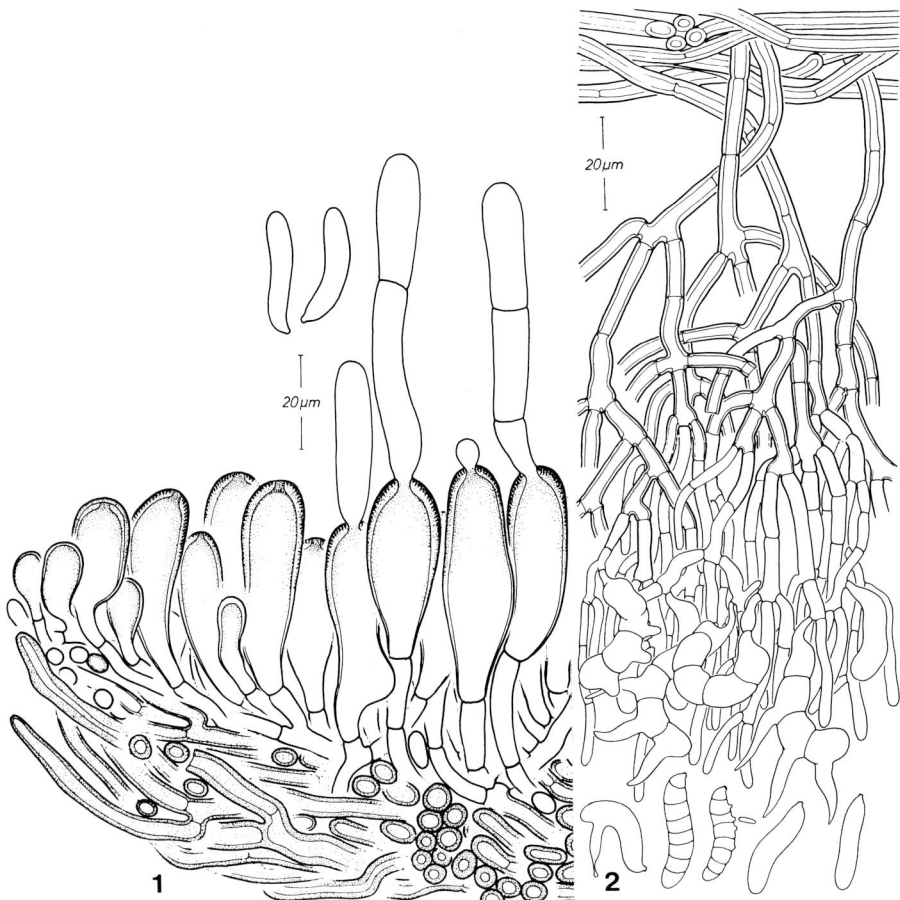

Fig. 1.5. Basidia of Septobasidiales. **1.5-1** *Uredinella coccidiophaga*. Portion of fruitbody with thick-walled probasidia, developing metabasidia, and basidiospores. **1.5-2** *Septobasidium albidum* Pat. Section of a fruitbody with mature basidia and basidiospores.

four-celled meiosporangia. Four genera of the Auriculariales with gastroid basidia are known: i.e., *Stilbum* (Juel, 1898), *Phleogena* (Shear and Dodge, 1925), *Pilacrella* (Schroeter, 1889), and *Hoehnelomyces* (Weese, 1920). The basidia of *Pilacrella solani* Cohn et Schroet. apud J. Schroet. (Fig. 1.3-4) are quite similar morphologically to those of the species of the auriculariaceous basidia mentioned above. The phylogeny of these taxa can possibly be traced by host specificity and basidiocarp morphology. The species of *Herpobasidium* are corticioid. The species of *Eocronartium* and the Indomalasian parasite on palms *Paraphelaria amboinensis* Corner (Corner, 1966a) form clavarioid basidiocarps; the gastroid genera (see above) have stalked and capitate basidiocarps.

Tremellales

Auriculariaceous and tremellaceous basidia are typologically quite different. In reality, however, the species of *Patouillardina* possess meiosporangia with irregular septations (Fig. 1.6), and are morphologically intermediate between those of these orders (Martin, 1935).

The meiosporangium of *Patouillardina* may be interpreted either as a relic of a basidium with variable septation or as a starting point in the evolution of the sphaeropedunculate basidium within the Tremellales. The balloon-shaped apical part of the basidium of *Patouillardina* is separated secondarily from the basal stalk by crossing longitudinal septa. As described above in the Auriculariales, this group of Tremellales characterized by sphaeropedunculate basidia (Fig. 1.7-1) apparently has evolved a series of taxa with increasingly complex fruiting bodies. The species of *Myxarium* (Fig. 1.7-3) develop basidiocarps that are corticioid or minutely pustulate in the initial stages of development. Isolated teeth are formed in *Stypella* (Fig. 1.7-2), and the fruiting bodies of *Protodontia* (Fig. 1.7-4) have an odontoid hymenial configuration. Stalked hydnoid species are referred to *Pseudohydnum,* the spathulate species with smooth hymenia to *Tremiscus*. While no agaricoid, lamellate representatives are known among these species with sphaeropedunculate basidia, minutely stalked and capitate species with sphaeropedunculate basidia represent the gastroid organization of *Hyaloria* (Möller, 1895; Fig. 1.7-5).

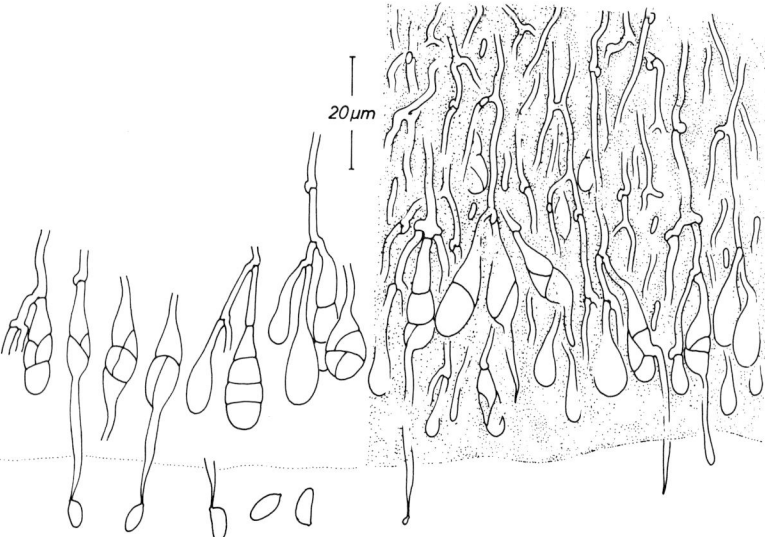

Fig. 1.6. *Patouillardina* sp. A section of the hymenium showing various stages of basidial ontogeny.

Significance of the Morphology of the Basidium

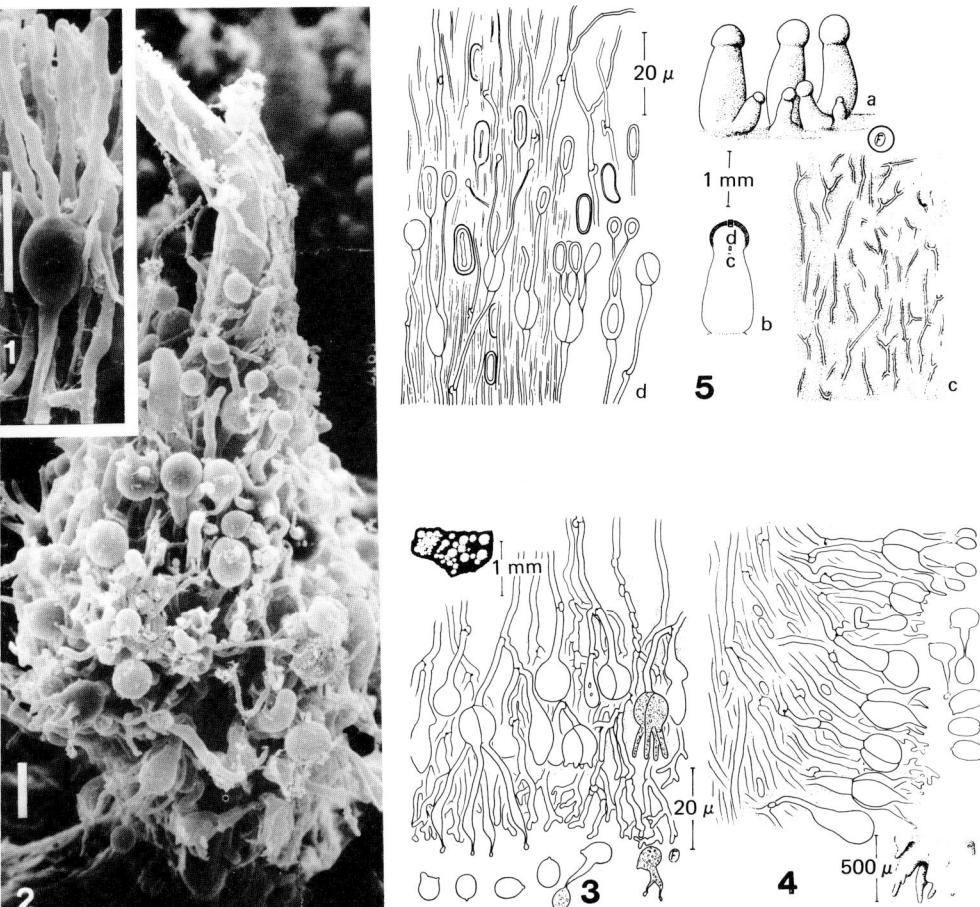

Fig. 1.7. Sphaeropedunculate basidia of different species of the Tremellaceae (Tremellales). **1.7-1** *Stypella papillata* Möller. Scanning electron micrograph of a basidium; bar = 10 μm. **1.7-2** *S. papillata*. Scanning electron micrograph of a fruitbody with basidia and basidiospores; bar = 10 μm. **1.7-3** *Myxarium* sp. Fruitbodies and part of the hymenium. **1.7-4** *Protodontia uda* v. Höhnel. Fruitbody and part of the hymenium with young and old basidia and basidiospores. **1.7-5** *Hyaloria pilacre* Möller. **a** Fruitbodies. **b** Longitudinal section of a single fruitbody. **c** Detail of the fruitbody stipe. **d** Part of the hymenium with immersed, gastroid basidia, basidiospores, and sterile hyphae.

Thus, based on comparative morphological studies of the basidium, the phylogeny of auriculariaceous and tremellaceous fungi could be interpreted in the manner diagrammed in Fig. 1.8. The irregularly septate meiosporangia of a new tropical species similar to those of *Patouillardina* and those of *P. cinerea* Bres. suggests a relatively close relationship between the Auriculariales and the Tremellales.

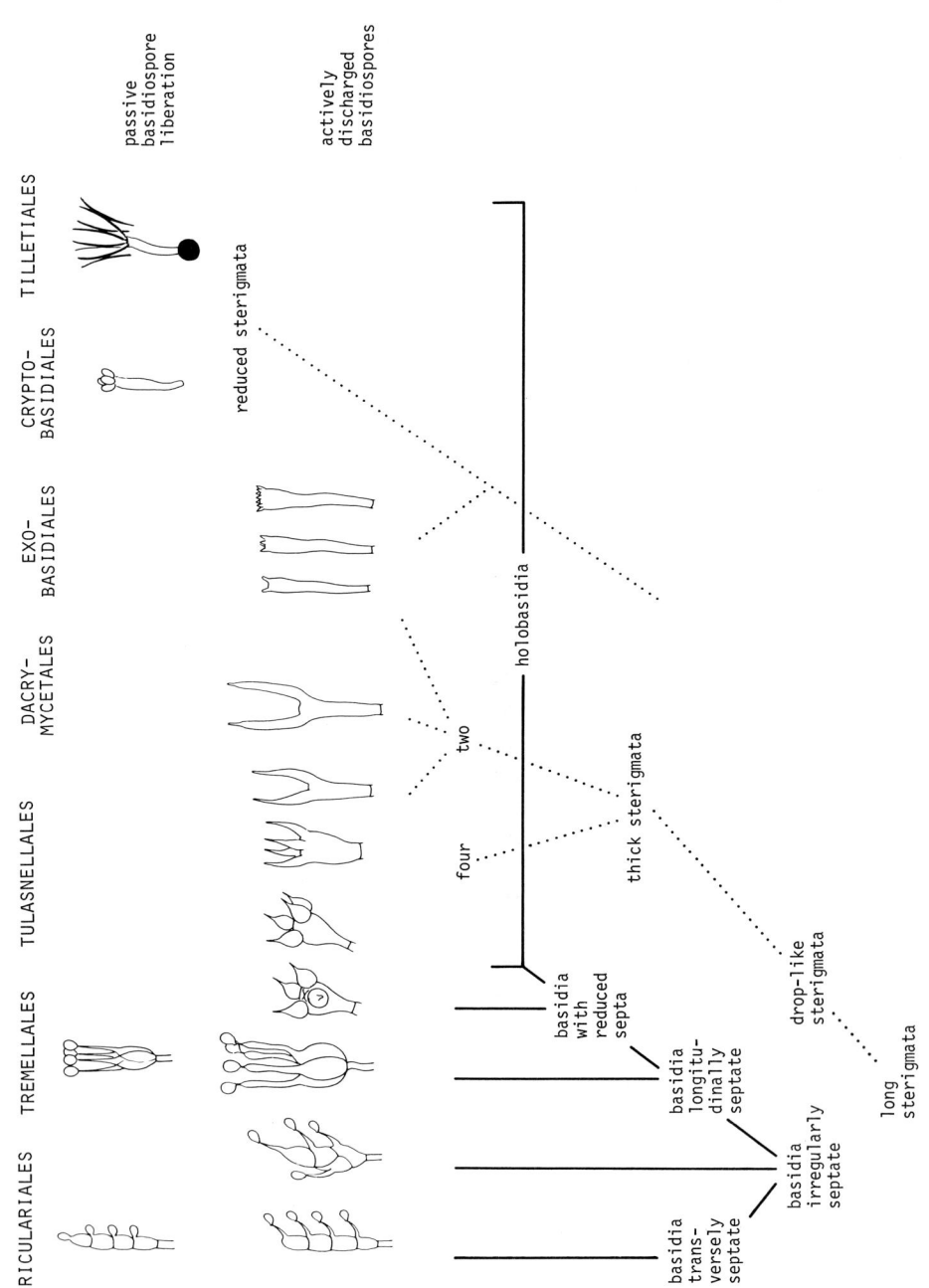

Holobasidiate Taxa

The boundary between the Tremellales and the groups with holobasidia is not sharply defined. There are two possible connections: (i) between the Tulasnellales and the Tremellales or (ii) between some Aphyllophorales and the Tremellales.

Pseudotulasnella (Lowy, 1964), with partial septations in the apical regions of the basidium, may represent a relic basidial type in the evolution of the holobasidium. Similar examples of partially septate basidia are known in the Tremellales (Fig. 1.9-1).

Along this apparent line of evolution of the holobasidium (Fig. 1.8), considerable variation in the structure of the sterigma can be observed. In general, deeply immersed basidia develop relatively long sterigmata, whereas superficially situated meiosporangia form relatively short sterigmata. The droplike sterigmata of the species of *Tulasnella* (Fig. 1.9-3) are not totally unlike the swollen sterigmata in *Ceratobasidium* (Fig. 1.9-4). The basidia of the species of *Ypsilonidium* (Fig. 1.9-5), with two sterigmata, superficially resemble the basidia of the Dacrymycetales (Fig. 1.9-7); however, this similarity might not represent a phylogenetic relationship. *Cerinomyces* (Fig. 1.9-6) probably does not link the Dacrymycetales with the Aphyllophorales.

The Dacrymycetales represent another one of the striking examples of fruit-body evolution beginning with corticioid (e.g., *Cerinomyces*) and pustulate (e.g., *Dacrymyces*) species and terminating, so far as is known today, with clavarioid (e.g., *Calocera*) taxa. All representatives of the Dacrymycetales are remarkably uniform in basidial morphology and ontogeny.

The meiosporangia of the Exobasidiales (Fig. 1.10-1) seem to differ totally from those of the Dacrymycetales. Nevertheless, species in *Exobasidium, Exobasidiellum, Brachybasidium* (Fig. 1.10-3), and *Dicellomyces* (Olive, 1945) develop forked basidia; however, the sterigmata of these meiosporangia are strongly reduced.

It is uncertain whether or not there are any connections from these holobasidiate taxa to the other parasitic groups on higher plants, such as the Cryptobasidiales (Malençon, 1953; Oberwinkler, 1977) (Fig. 1.10-5, 1.10-6) and the Tilletiales s. str. (Fig. 1.10-7). Species of both the Cryptobasidiales and the Tilletiales have gastroid basidia with strongly reduced sterigmata and basidiospores that seem to be passively detached (Figs. 1.10-5, 1.10-6, 1.10-7).

It is possible that the evolution of the thick-walled probasidium (Fig. 1.2) in the Uredinales and related taxa (e.g., Septobasidiales) could have also resulted in the thick-walled probasidium of the gastroid Ustilaginales. There are also several examples of thick-walled probasidia in the parasitic holobasidiate Heterobasidiomycetes, e.g., Exobasidiales and Cryptobasidiales. There are, however, no obvious connections between these taxa and the Tilletiales.

◀ **Fig. 1.8.** Basidial types of some Heterobasidiomycetes. The characteristic basidial types are linked by intermediate types (continuous line). The variations and the postulated phylogeny of sterigmata is shown (dotted line). Further, the development of the gastroid basidium is indicated.

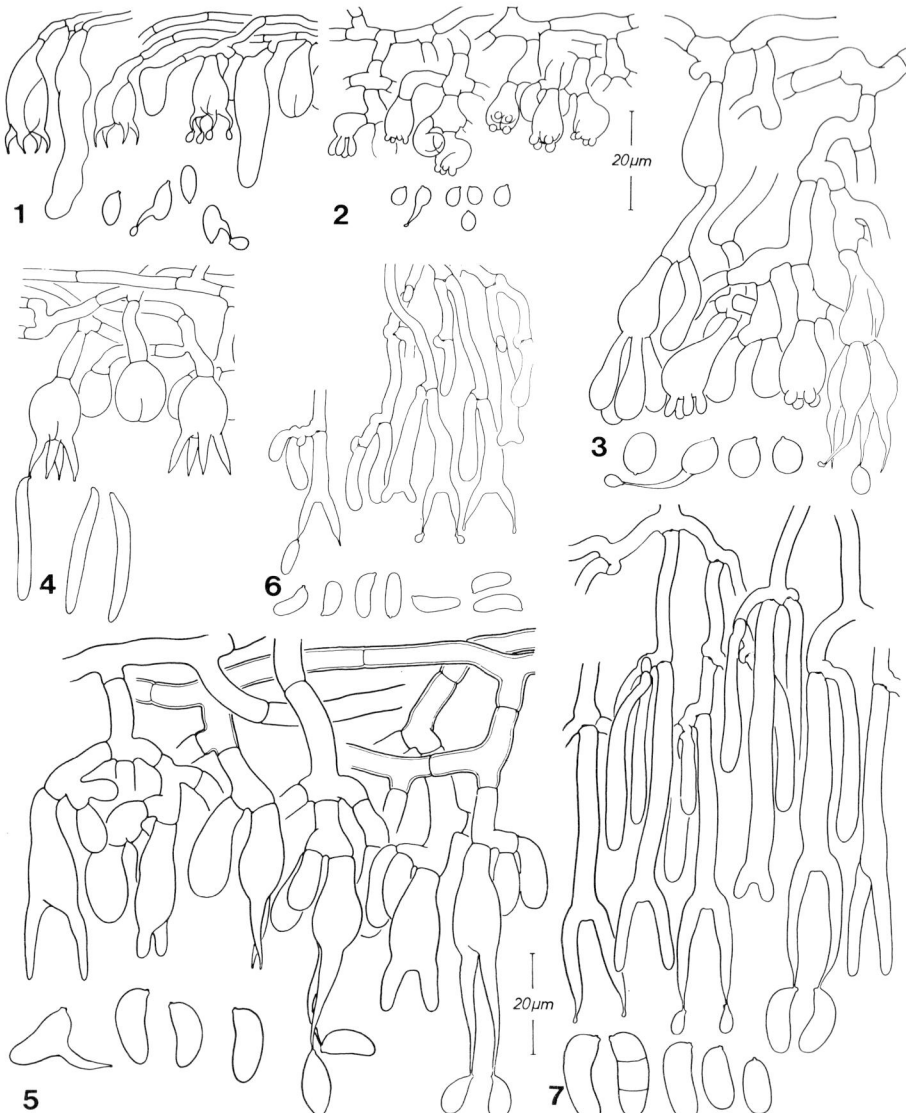

Fig. 1.9. Some representative and intermediate basidial types of the Heterobasidiomycetes. **1.9-1** *Sebacina* sp. (s. l.). An undescribed species with partially septate basidia. **1.9-2** *Pseudotulasnella guatemalensis* Lowy. Portion of the fruitbody with basidia and basidiospores. **1.9-3** *Tulasnella violea* (Quél.) Bourd. et Galz. Section through the fruitbody with basidia and basidiospores. **1.9-4** *Ceratobasidium calosporum* D. P. Rogers. Section through the fruitbody with basidia and basidiospores. **1.9-5** *Ypsilonidium sterigmaticum* (Bourd.) Donk. Portion of the fruitbody with basida and basidiospores. **1.9-6** *Cerinomyces crustulinus* (Bourd. et Galz.) Martin. Part of the fruitbody with basidia and basidiospores. **1.9-7** *Dacrymyces punctiformis* Neuh. A portion of the hymenium with basidia and basidiospores.

Fig. 1.10. Representative basidia of the Exobasidiales, Cryptobasidiales, and Tilletiales. **1.10-1** *Exobasidium oxycocci* Rostr. A cluster of basidia, one mature. **1.10-2** *Exobasidiellum graminicola* (Bres.) Donk. Basidia and basidiospores. **1.10-3** *Brachybasidium pinangae* (Rac.) Gäum. Section of the fruitbody with basidia. **1.10-4** *Digitatispora marina*. A cluster of **a** fruitbodies, **b** section of a fruitbody, and **c** a part of the hymenium with basidia and basidiospores. **1.10-5** *Microstroma juglandis* Bereng. A cluster of basidia with basidiospores with some unattached spores with yeast-like germination. **1.10-6** *Coniodictyum chevalieri* Har. et Pat. Basidia and basidiospores. **1.10-7** *Tilletia caries* (DC.) Tul. Smut spore, basidium and sporidia.

A basidiomycete of vague phylogenetic position is *Digitatispora marina* Doquet (Doquet, 1962) (Fig. 1.10-4). The branched spores of this species may reflect the adaptation to the marine habitat of *D. marina*. At the present time, I am inclined to believe, on the basis of basidial morphology, that this species should be classified at a gastroid level near the Dacrymycetales.

Basidiomycete Yeasts and Yeast-like Taxa

Some species within the Heterobasidiomycetes have affinities with yeast-like somatic phases (Fig. 1.11). An interpretation of sexual reproduction should aid in clarifying the natural relationships of these taxa. *Rhodosporidium* (Banno, 1967; Fell et al., 1970) and *Leucosporidium* (Fell et al., 1969; Fell and Phaff, 1970) form basidia similar to the basidia of the Ustilaginales (Fig. 1.11). On the other hand, *Aessosporon* (van der Walt, 1970) and *Sporidiobolus* (Nyland, 1949; Laffin and Cutter, 1959a, 1959b; Phaff, 1970; Bandoni et al., 1971) form tilletiaceous basidia (Fig. 1.11). The morphology of the meiosporangia of *Filobasidium* (Olive, 1968) and *Filobasidiella* (Kwon-Chung, 1975, 1976a, 1976b) suggests a possible relationship to these taxa with the Cryptococcales; however, I am not yet prepared to offer concrete proposals on the phylogeny of this group without additional information.

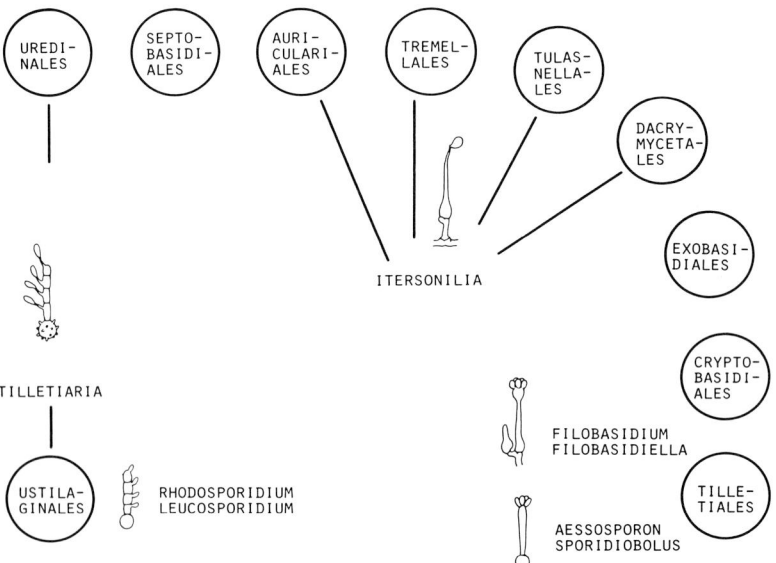

Fig. 1.11. Orders of Heterobasidiomycetes with possible connections to genera with yeast-like species.

More details, however, are available to support discussions of the relationships of such species as *Itersonilia perplexans* Derx (Derx, 1948; Olive, 1952). Basidia with a single sterigma, such as those formed by *I. perplexans* (Fig. 1.12) do occur in other basidiomycete taxa, which are, however, widely separated taxonomically. Of particular interest are the single sterigmate basidia of the Auriculariales, Tremellales, Tulasnellales, and Dacrymycetales. The single-spored basidia of *Platygloea unispora* Olive (Olive, 1944, 1947) and *Dacrymyces ovisporus* Bref. (Brefeld, 1888), which I find are difficult to separate (see Bandoni, 1963), lack the characteristic probasidial swelling of *Itersonilia perplexans* (Fig. 1.12). I know of one collection of an undescribed species of *Tulasnella* that forms few- to single-spored basidia, which are quite different morphologically from those of *I. perplexans* (Fig. 1.12). Single-spored basidia of tremellaceous species are quite similar morphologically to the sporogenous cells of *I. perplexans*.

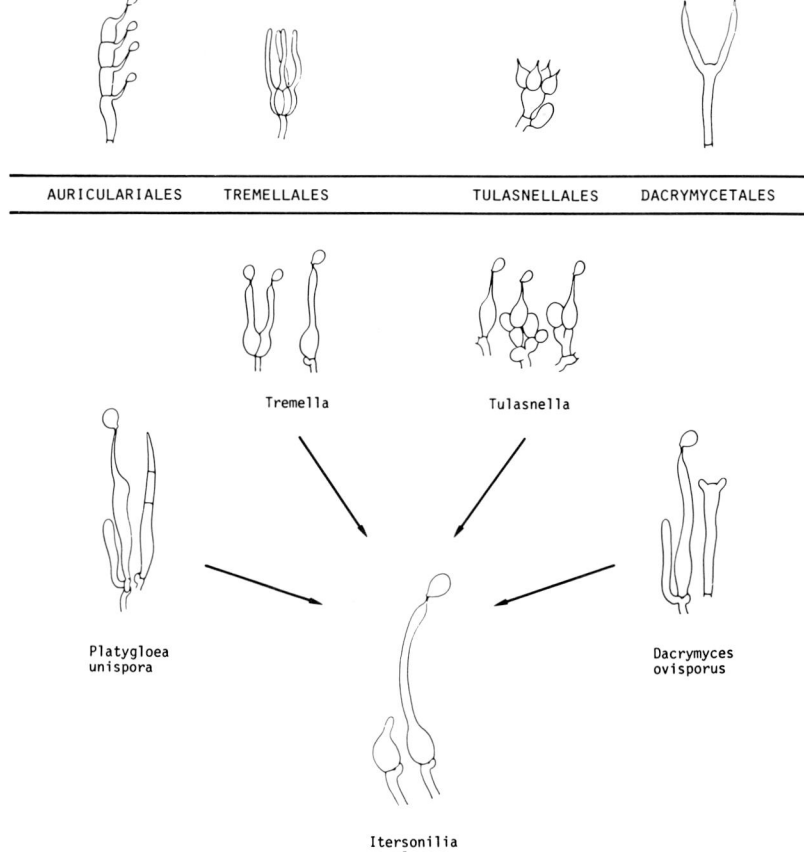

Fig. 1.12. Comparison of the basidia of *Itersonilia perplexans* with one-spored meiosporangia of other taxa of the Heterobasidiomycetes.

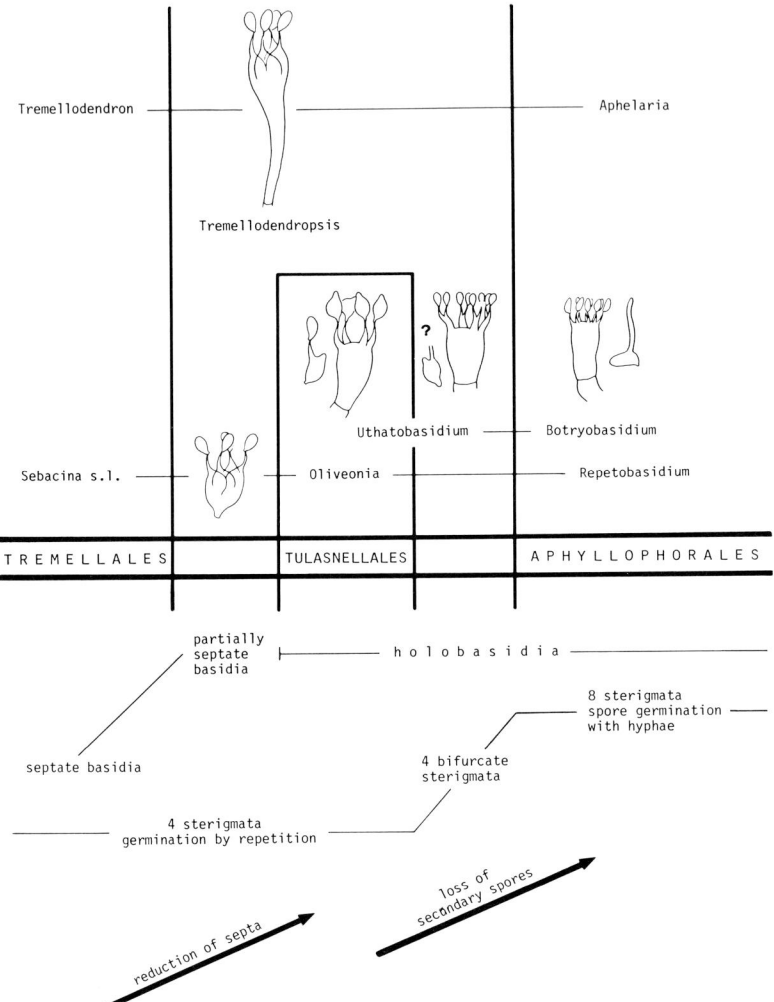

Fig. 1.13. The interrelationships between the Tremellales, Tulasnellales, and Aphyllophorales based on comparative basidial morphology. The primary evolutionary steps may have been the reduction of septa of the phragmobasidium and the loss of secondary spores of the heterobasidium.

Fig. 1.14. Basidial variation within the Homobasidiomycetes. The urnigera type of basidium in the species of *Sistotrema* intergrade successively towards the *Botryobasidium* type, the suburniform meiosporangium of *Hyphodontia,* and the basidium of the species of *Paullicorticium*. The pleurobasidia of the species of the Xenasmataceae can be connected to different basidial types in the Vuilleminiaceae (see also Fig. 1.15).

Significance of the Morphology of the Basidium

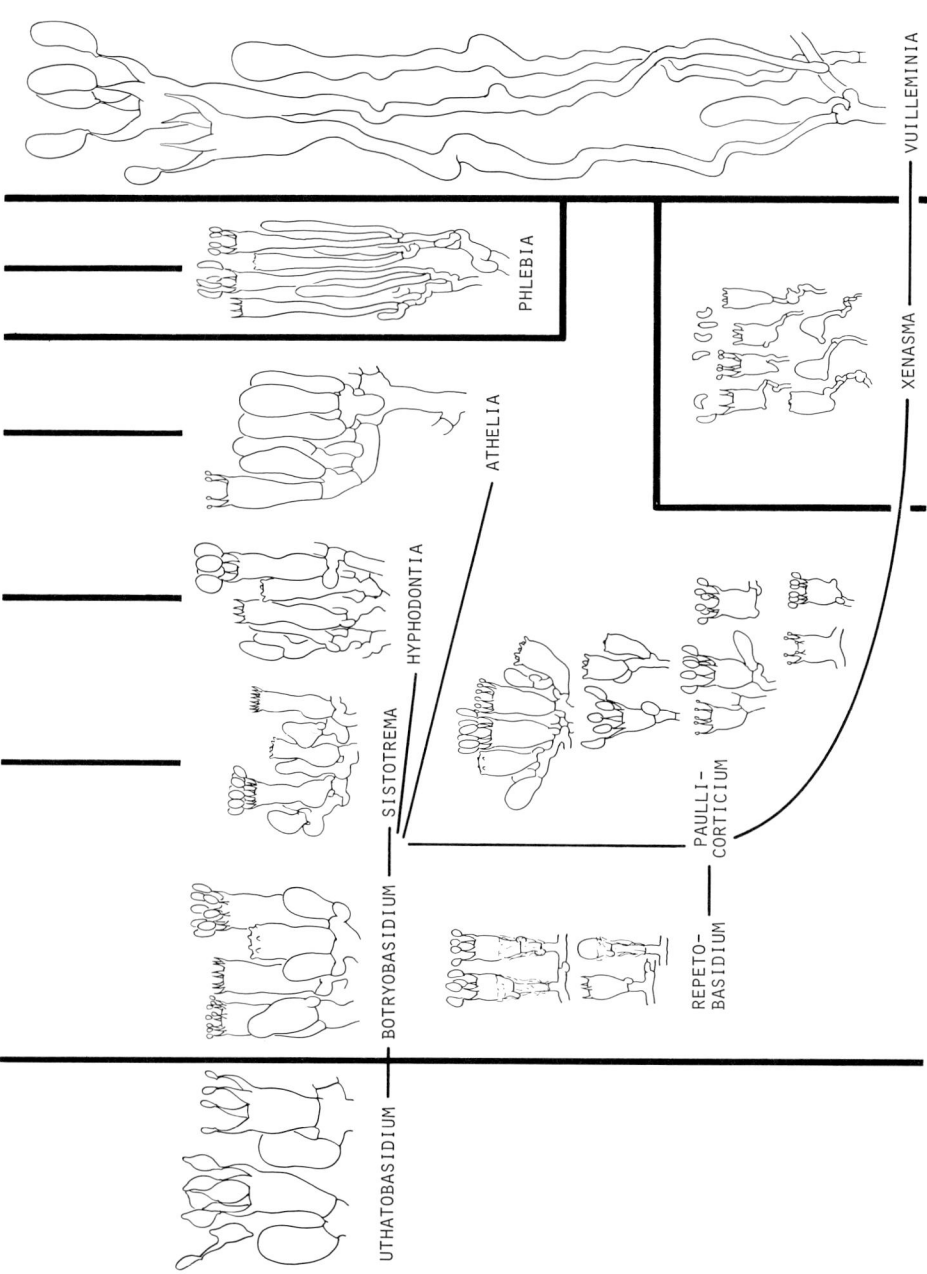

Homobasidiomycetes

The existence of species with partially septate basidia (Oberwinkler, 1972) makes it possible to postulate the evolution of the Homobasidiomycetes from several phragmo- and holobasidial taxa (Fig. 1.13). The possibility of the holobasidium evolving via *Pseudotulasnella* from the Tulasnellales has been discussed above (see Holobasidiate Taxa). *Tremellodendropsis* (Crawford, 1954; Corner, 1966b) with apically septate basidia could represent an evolutionary stage between *Tremellodendron* of the Tremellales and *Aphelaria*-like clavarioid species of the Aphyllophorales.

The evolution of the Homobasidiomycetes from a phragmo- and/or holobasidial ancestral form was probably accompanied by the loss of the capacity to form secondary spores. In the Ceratobasidiaceae and in *Botryobasidium* (Fig. 1.14) this loss might be explained by the development of secondary spores directly on the basidium (Oberwinkler, 1965). Supernumerous sterigmata in *Botryobasidium* could have resulted from such evolution, which shows remarkable similarity with basidia with bifurcate sterigmata (Fig. 1.13).

Such an interpretation is not reasonable for *Oliveonia* and *Repetobasidium*, which are so similar that a differentiation between the two genera is problematical without information on the type of spore germination. Within the corticiaceous Aphyllophorales, a gradual transition in basidial morphology is known (Fig. 1.14). This transition probably reflects close relationships as well as a direction of evolution towards a uniform meiosporangium. The main basidial types (Fig. 1.14) are urniform (e.g., *Sistotrema*), suburniform (e.g., *Hyphoderma, Hyphodontia*), pleurobasidial (e.g., *Xenasma*), as well as long and contorted (e.g., *Vuilleminia*). Typologically a number of intermediate forms (e.g., *Paullicorticium* species) tie together the main types (Fig. 1.14), as for example some *Paullicorticium* species (Oberwinkler, 1965).

Pleurobasidial meiosporangia develop laterally on creeping hyphae in several taxa (Fig. 1.15); this is very characteristic for several species with gelatinous fruitbodies. There is some evidence (Oberwinkler, 1965) that phylogenetic differentiation led to the elongated basidia of the Vuilleminiaceae (Fig. 1.16). Although the majority of species differentiating organized hymenia develop basidial clusters with terminal meiosporangia (Fig. 1.15), pleurobasidia can be found in some such groups, especially in superficially creeping generative hyphae at the margins of the fruiting bodies. Thus, the pleurobasidium could be interpreted as an ancestral type from which the elongated, terminal basidia have evolved.

As I have described above (see Auriculariales, Tremellales, and Holobasidiate

Fig. 1.15. Basidial evolution within the Homobasidiomycetes. Typical pleurobasidia of the Xenasmataceae develop laterally in a gelatinous matrix. The elongated basidia of the Vuilleminiaceae, which are also formed in a gelatinous matrix, seem to be related morphologically to the pleurobasidium. The basidia of other taxa, such as the Corticiaceae s. l., Agaricales, and gasteromycetes develop terminally; however, in some taxa at the periphery of the fruitbody, creeping hyphae can form basidia laterally as in the Xenasmataceae.

Significance of the Morphology of the Basidium

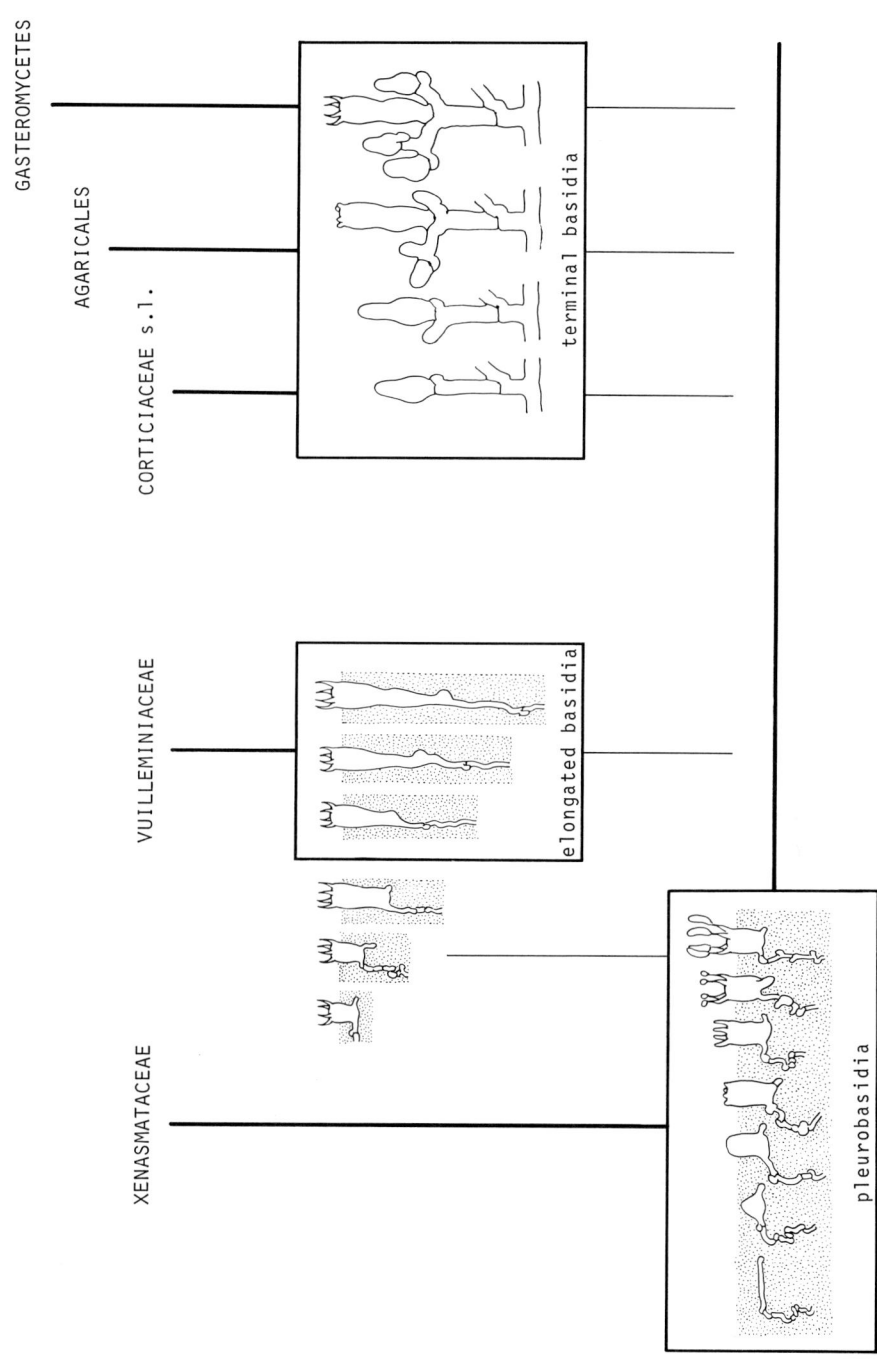

Taxa), there is phylogenetic diversification of the fruiting bodies of the Auriculariales, Tremellales, and Dacrymycetales. This can also be documented in the Homobasidiomycetes (Fig. 1.16; Oberwinkler, 1979), e.g., in the *Sistotrema, Hyphodontia, Athelia,* and *Phlebia* groups.

The presence of urniform basidia in both taxa suggests a phylogenetic relationship between the corticioid *Sistotrema* and the clavarioid *Multiclavula* (Fig. 1.16-1a). Some species of *Phlebia* (Fig. 1.16-4a), *Merulius* s. str., *Porodisculus,* and *Panellus* (Fig. 1.16-4b) form characteristic meruliaceous basidia; however, there are few other correlating characters, which makes it difficult to postulate an evolutionary connection between these taxa. The species of *Clavaria* s. str. (Fig. 1.16-2a) and *Humidicutis* (Fig. 1.16-2b) form identical basidia with characteristic sub-basidial looplike clamps, but there is insufficient information to warrant speculation on the phylogeny of this basidial form. Clavate basidia without specific characteristics, as present in the species of *Athelia* (Fig. 1.16-3a), are found, for example, in *Phanerochaete, Cyphellostereum, Stereophyllum, Hydnopolyporus, Albatrellus, Omphalina* (Fig. 1.16-3b), and *Tricholoma* species.

Within the Agaricales with uniform basidia, I am aware of two types of meiosporangia: non-inflating basidia (Fig. 1.17-1a, 1.17-1b) and those that strongly expand apically (Fig. 1.17-2a, 1.17-2b, 1.17-2c). Very often those species forming basidia that expand apically form basidiocarps in which the hyphae of the trama, subhymenium, and hymenium also show similar, secondary expansion. Possibly this feature is associated with the capacity of basidiocarps to undergo rapid expansion under optimum environmental conditions. It is likely that relationships do exist that connect taxa with non-inflating basidia and those with apically expanding basidia.

A high percentage of Aphyllophorales and most Agaricales bear basidia of uniform morphology. This reduces the value of comparative morphological studies of the basidia in the more complex taxa of these orders. Comparative studies are of some value in efforts to determine the relationships of the gasteromycetes with the Agaricales and Aphyllophorales (Fig. 1.18).

The remainder are the isolated groups of the gasteromycetes (Fig. 1.18). Most of these taxa have quite specific basidia (Fig. 1.18), e.g., Lycoperdales, Geastrales, Sclerodermatales, Melanogastrales, Gautieriales, Tulostomatales, Nidulariales, and Phallales (Oberwinkler, 1977). This basidial diversification in these orders can best be interpreted as the end result of long-separated evolution. I am not aware of any connections of these basidial types with those of other groups; therefore, I cannot deal with the phylogeny of the true gasteromycetes on the basis of comparative basidial morphology.

Fig. 1.16. Basidia of the Aphyllophorales and the Agaricales. **1.16-1** Urniform basidia of **a** *Sistotrema brinkmannii* (Bres.) J. Eriks. and of **b** *Multiclavula mucida* (Fr.) Petersen. **1.16-2** The basidia of **a** *Clavaria tenuipes* B. et Br. and **b** *Humidicutis marginatus* (Peck) Singer with loop-like, sub-basidial clamps. **1.16-3** The urniform type of holobasidia of **a** *Athelia epiphylla* Pers. and **b** *Omphalina lutealilacina* (Favre) Hendersen. **1.16-4** The meruliaceous type of basidia as exemplified by **a** *Phlebia radiata* Fr. and **b** *Panellus violaceofulvus* (Batsch ex Fr.) Singer.

Significance of the Morphology of the Basidium

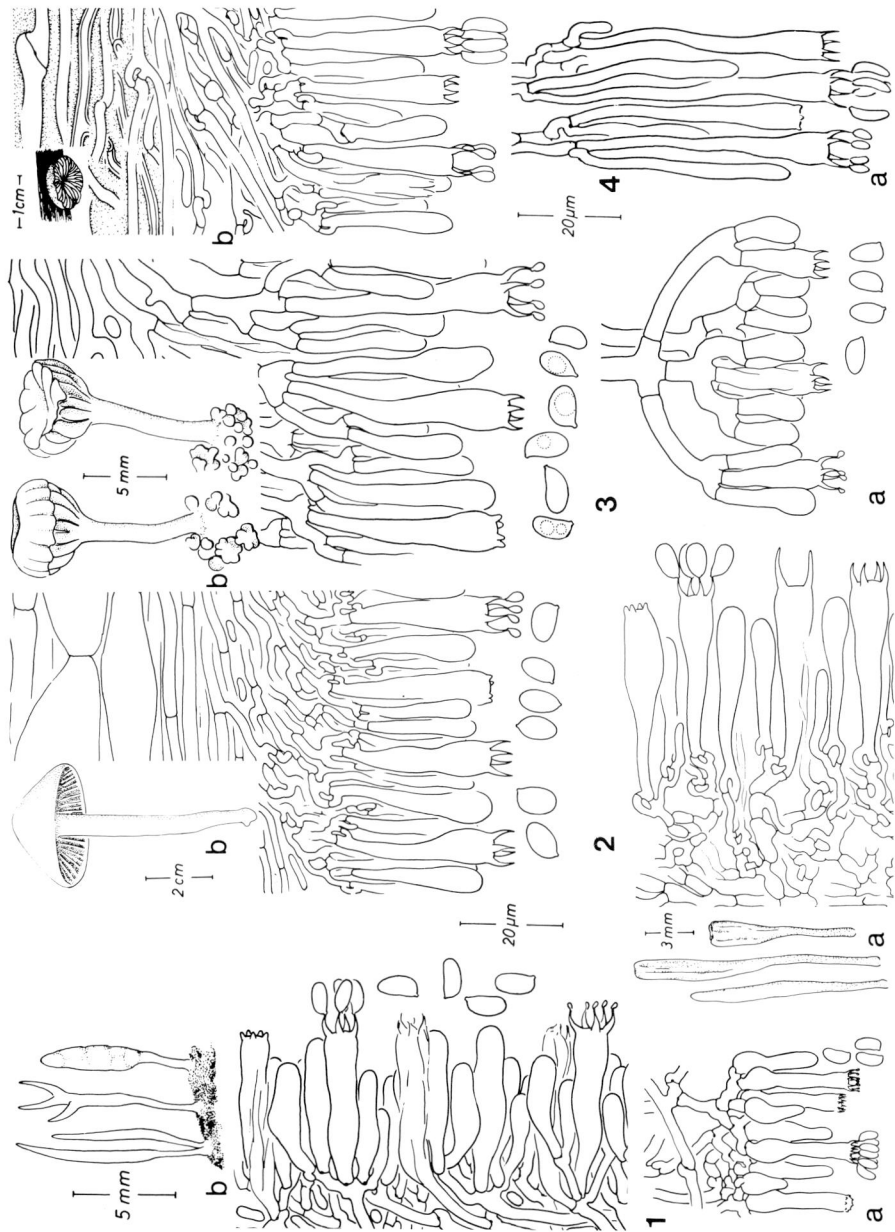

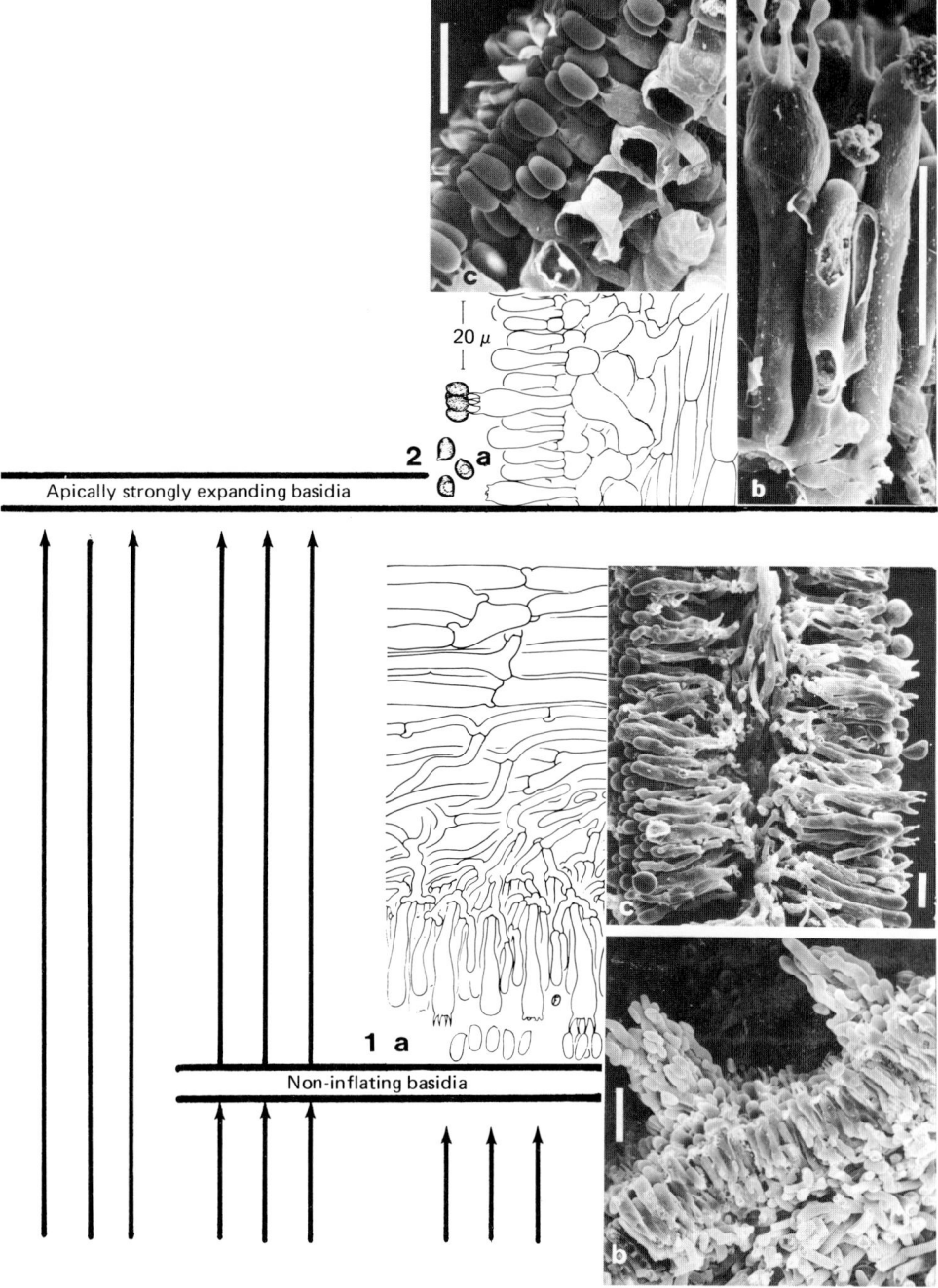

Fig. 1.17. Basidial types of the Agaricales. **1.17-1** The non-inflating basidia of **a** *Xeromphalina campanella* (Batsch ex Fr.) Maire, **b** *Lentinus tigrinus* (Bull. ex Fr.) Fr., and **c** *Mucidula mucida* (Fr.) Pat. **1.17-2** The strongly apically expanding basidia of **a** *Agaricus xanthoderma* Gen., **b** *Russula mairei* Singer, and **c** *Coprinus comatus* Fr.

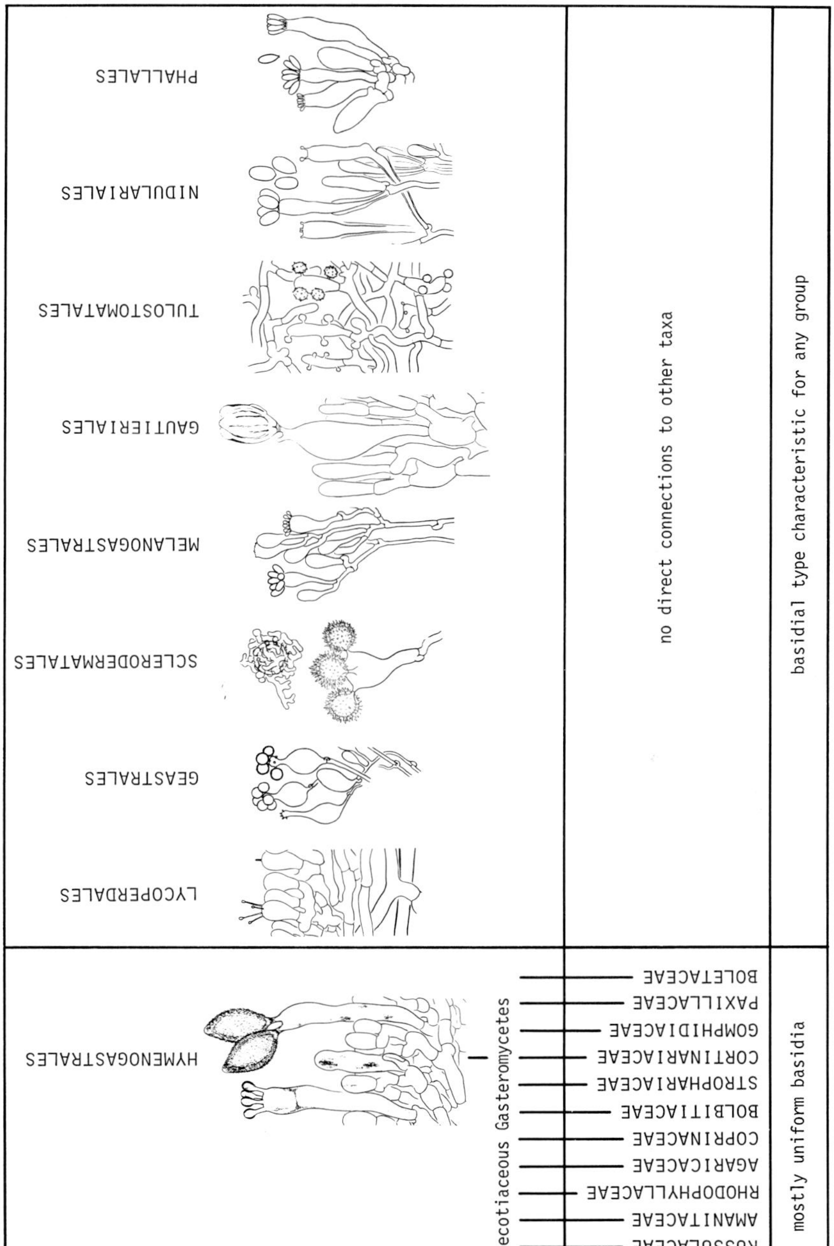

Fig. 1.18. Connections between the Agaricales and secotiaceous gasteromycetes, and the basidial types characteristic of the isolated groups of the gasteromycetes.

Acknowledgments. I am grateful to Dr. K. Wells for suggestions for improvement in the English text. Dr. R. J. Bandoni's personal communications, especially those concerning yeasts and species with yeast-like stages, have contributed to my understanding of these taxa. The cultures that he has made available to me have also been of essential value in my studies. I also wish to thank Mrs. H. Gminder and Dr. C. Sautter for their technical assistance in the use of the scanning electron microscope.

References

Bandoni, R. J.: *Dacrymyces ovisporus* from British Columbia. Mycologia *55*, 360–361 (1963).
Bandoni, R. J., Johri, B. N.: *Tilletiaria:* a new genus in the Ustilaginales. Can. J. Bot. *50*, 39–43 (1972).
Bandoni, R. J., Lobo, K. J., Brezden, S. A.: Conjugation and chlamydospores in *Sporobolomyces odorus*. Can. J. Bot. *49*, 683–686 (1971).
Banno, J.: Studies in the sexuality of *Rhodotorula*. J. Gen. Appl. Microbiol. *13*, 167–196 (1967).
Brefeld, O.: Untersuchungen aus dem Gesammtgebiete der Mykologie. 7. Basidiomyceten II. Protobasidiomyceten. Leipzig: Arthur Felix 1888, 178 pp. + 11 plates.
Corner, E. J. H.: *Paraphelaria*, a new genus of Auriculariaceae (Basidiomycetes). Persoonia *4*, 345–350 (1966a).
Corner, E. J. H.: The clavarioid complex of *Aphelaria* and *Tremellodendropsis*. Trans. Br. Mycol. Soc. *49*, 205–211 (1966b).
Couch, J. N.: A new fungus intermediate between the rusts and *Septobasidium*. Mycologia *29*, 665–673 (1937).
Crawford, D. A.: Studies on New Zealand Clavariaceae. I. Trans. R. Soc. N.Z. Bot. *82*, 617–631 (1954).
Derx, H. G.: *Itersonilia*, nouveau genre de Sporobolomycètes à mycélium bouclé. Bull. Bot. Gard. Buitenzorg, Ser. III, *17*, 465–472 (1948).
Doguet, G.: *Digitatispora marina*, n. g., n. sp., Basidiomycète marin. C. R. Hebd. Séances Acad. Sci. *254*, 4336–4338 (1962).
Donk, M. A.: A conspectus of the families of Aphyllorphorales. Persoonia *3*, 199–324 (1964).
Fell, J. W., Phaff, H. J.: Genus 1. *Leucosporidium* Fell, Statzell, Hunter et Phaff. In: The Yeasts. Lodder, J. (ed.). Amsterdam: North-Holland Publishing Co. 1970, pp. 776–802.
Fell, J. W., Phaff, H. J., Newell, S. Y.: Genus 2. *Rhodosporidium* Banno. In: The Yeasts. Lodder, J. (ed.). Amsterdam: North-Holland Publishing Co. 1970, pp. 803–814.
Fell, J. W., Statzell, A. C., Hunter, J. L., Phaff, H. J.: *Leucosporidium* gen. n., the heterobasidiomycetous stage of several yeasts of the genus *Candida*. Antonie van Leeuwenhoek J. Microbiol. Serol. *35*, 433–462 (1969).
Gäumann, E.: Über die Entwicklungsgeschichte von *Jola javensis* Pat. Ann. Mycol. *20*, 272–289 (1922).
Gould, C. J.: The parasitism of *Glomerularia lonicerae* (Pk.) D. and H. in *Lonicera* species. Iowa St. Coll. J. Sci. *19*, 301–331 (1945).
Jackson, H. S.: The nuclear cycle in *Herpobasidium filicinum* with a discussion of the significance of homothallism in the Basidiomycetes. Mycologia *27*, 553–572 (1935).

Juel, H. O.: *Stilbum vulgare* Tode. Ein bisher verkannter Basidiomycet. Bih. K. svenska Vet. Akad. Handl. (III) *24*, (9), 1–15 (1898).

Kwon-Chung, K. J.: Description of a new genus, *Filobasidiella*, the perfect state of *Cryptococcus neoformans*. Mycologia *67*, 1197–1200 (1975).

Kwon-Chung, K. J.: Morphogenesis of *Filobasidiella neoformans*, the sexual state of *Cryptococcus neoformans*. Mycologia *68*, 821–833 (1976a).

Kwon-Chung, K. J.: A new species of Filobasidiella, the sexual state of *Cryptococcus neoformans* B and C serotypes. Mycologia *68*, 942–946 (1976b).

Laffin, R. J., Cutter, V. M., Jr.: Investigations on the life cycle of *Sporidiobolus johnsonii*. I. Irradation and cytological studies. J. Elisha Mitchell Sci. Soc. *75*, 89–96 (1959a).

Laffin, R. J., Cutter, V. M., Jr.: Investigations on the life cycle of *Sporidiobolus johnsonii*. II. Mutants and micromanipulation. J. Elisha Mitchell Sci. Soc. *75*, 97–100 (1959b).

Lowy, B.: A new genus of the Tulasnellaceae. Mycologia *56*, 696–700 (1964).

Malençon, G.: Le *Coniodyctium chevalieri* Har. et Pat., sa nature et ses affinités. Bull. Soc. Myc. Fr. *69*, 77–100 (1953).

Martin, G. W.: *Atractobasidium*, a new genus of the Tremellaceae. Bull. Torrey Bot. Club *62*, 339–343 (1935).

Möller, A.: Protobasidiomyceten. Bot. Mitth. Trop. *8*, 1–179 (1895).

Nyland, G.: Studies on some unusual Heterobasidiomycetes from Washington State. Mycologia *41*, 686–701 (1949).

Oberwinkler, F.: Primitive Basidiomyceten. Sydowia Ann. Mycol. *1*, 1–72 (1965).

Oberwinkler, F.: The relationships between the Tremellales and the Aphyllophorales. Persoonia *7*, 1–16 (1972).

Oberwinkler, F.: Das neue System der Basidiomyceten. In: Beiträge zur Biologie der niederen Pflanzen. Frey, H., Hurka, H. Oberwinkler, F. (eds.). Stuttgart: G. Fischer 1977, pp. 59–105.

Oberwinkler, F.: Beziehung aphyllophoraler zu agaricalen Basidiomyceten. Beihefte zur Sydowia, Ann. Mycol. Ser. II. *8*, 276–289 (1979).

Olive, L. S.: New or rare Heterobasidiomycetes from North Carolina. I. J. Elisha Mitchell Sci. Soc. *60*, 17–26 (1944).

Olive, L. S.: A new *Dacrymyces*-like parasite of *Arundinaria*. Mycologia *37*, 543–552 (1945).

Olive, L. S.: Notes on the Tremellales of Georgia. Mycologia *39*, 90–108 (1947).

Olive, L. S.: Studies on the morphology and cytology of *Itersonilia perplexans* Derx. Bull. Torrey Bot. Club *79*, 126–138 (1952).

Olive, L. S.: An unusual new heterobasidiomycete with *Tilletia*-like basidia. J. Elisha Mitchell Sci. Soc. *84*, 261–266 (1968).

Phaff, H. J.: *Sporidiobolus* Nyland. In: The Yeasts. Lodder, J. (ed.). Amsterdam: North-Holland Publishing Co. 1970, pp. 822–830.

Schroeter, J.: Pilze. In: Kryptogamen-Flora von Schlesien: im Namen der Schlesischen Gesellschaft für vaterländische Cultur herausgegeben *3* (1). Cohn, F. (ed.). Breslau: J. U. Kern's Verlag 1889, pp. 1–814.

Shear, C. L., Dodge, B. O.: The life history of *Pilacre faginea* (Fr.) B. & Br. J. Agr. Res. *30*, 407–417 (1925).

van der Walt, J. P.: The perfect and imperfect states of *Sporobolomyces salmonicolor*. Antonie van Leeuwenhoek J. Microbiol. Serol. *36*, 49–55 (1970).

Weese, J.: Beitrag zur Morphologie und Systematik einiger Auriculariineengattungen. Ber. Dtsch. Bot. Ges. *7*, 512–519 (1920).

Chapter 2
Ultrastructure and Cytochemistry of Basidial and Basidiospore Development

DAVID J. MCLAUGHLIN

Introduction

Although the basidium has been recognized as a distinctive type of reproductive cell since the early nineteenth century, there has been relatively little interest until recently in cytoplasmic events in its development. Corner's (1948) detailed light-microscopic account of non-nuclear aspects of basidial development provided a major impetus and a theoretical framework for future studies. Ultrastructural investigation of the basidium began in the 1960s. Most of this research has dealt with the holobasidium but only the Agaricales and Aphyllophorales have been carefully examined. The cytoplasmic events in holobasidial development even in *Coprinus*, which is perhaps the best studied genus (McLaughlin, 1977b), are not yet fully known. The number of publications on phragmobasidial ultrastructure has increased considerably in recent years, but a comprehensive account of the cytoplasmic events in phragmobasidial development is still lacking.

This chapter will be limited to a discussion of changes in the cytoplasm and walls during basidial development, exclusive of nuclear division. It does not provide a complete account of all papers dealing with the ultrastructure of basidia and basidiospores. Papers that have had as their primary goal the gathering of taxonomic information or have contributed little to understanding basidial development have been omitted.

Ultrastructural studies of non-nuclear events in basidial development have several objectives. They provide a structural basis for understanding basidial morphogenesis and especially the features unique to the Basidiomycotina, i.e., the formation of violently discharged basidiospores on sterigmata. These studies may allow us to determine if there is a common ultrastructural pattern for basidial development which is applicable to both holobasidia and phragmobasidia. Ultrastructural studies may also lead to a better understanding of the evolutionary relationships between holobasidiate and phragmobasidiate groups.

The holobasidium, especially that of *Coprinus cinereus* (Schaeff. ex. Fr.) S. F. Gray, will be emphasized in this account. About 3 days are required to complete basidial development in *C. cinereus*. Basidiospores are released in the early morning of a day designated day 0. Earlier events in basidial development can be measured backward from the time of basidiospore release. Basidium initiation occurs at about −3 days. Karyogamy occurs at −2 days in early evening. Meiosis begins at about midnight, −2 days, and is completed at −1 day in late morning. Sterigmal initiation begins about 1 h after completion of meiosis and basidiospore initiation begins after an additional 1 h. Timing of these events with respect to the day/night cycle is not absolute, but varies between cultures by only a few hours.

Basidial Initiation

The holobasidium usually arises as the terminal cell of a hypha or hyphal branch. Other methods of basidial origin are possible but less common (Sundberg, 1977). Of the seven stages into which Corner (1948) divided holobasidial development, three stages dealt with the basidium from its initiation until it reached full size. At "inception", the first stage, the basidium is an enlarging hyphal tip. In the second stage, "charging", it enlarges further while filling with cytoplasm. The basidium may attain full size with or without entering stage 3, "initial vacuolation" (Sundberg, 1977).

An examination of ultrastructural aspects of basidial initiation in *Coprinus cinereus* revealed that the cytoplasm was relatively simple in young basidia (−3 days; Fig. 2.1) compared with postmeiotic basidia (Fig. 2.16). The basidium at the early prefusion stage contained numerous ribosomes, a few vacuoles, mitochondria, and limited endoplasmic reticulum. It has been generally assumed that the hypha which forms the basidium must show a cytoplasmic organization like that of a growing vegetative hyphal apex (Sundberg, 1977). Thus far our results indicate little resemblance to tip growth in hyphae (Grove and Bracker, 1970). The absence of vesicles at the apex of young basidia may indicate a slow growth rate and be comparable to the apex of rust hyphae, which also lacks typical hyphal tip organization (Coffey, 1975). An apical accumulation of vesicles suggesting tip growth has been observed in the primary and secondary (i.e., the second promycelium formed on a teliospore) promycelia of *Ustilago maydis* (DC) Corda (Ramberg, 1979) and in promycelia of *Puccinia malvacearum* Bert. (McLaughlin and O'Donnell, unpubl. results). The relatively simple cytoplasm (with few types of organelles) of young basidia of *C. cinereus* seems also to be typical of other holobasidia (Wells, 1965; Beckett et al., 1974, p. 205), as well as of some phragmobasidia (Wells, 1964b).

Comparison of holobasidial development with species that form teliospores presents a problem. In the latter, development occurs in two stages. The probasidium passes through a dormant phase before formation of the metabasidium. In members of the Uredinales which have been studied ultastructurally, probasidia are initiated as sympodioconidia (Littlefield and Heath, 1979) and early development is indistinguishable from that of urediospores (Harder, 1977). Mature teliospores or those at early germination had dense cytoplasm which was difficult to study by conventional thin-sectioning techniques. They usually contained relatively few types of organelles but had numerous lipid droplets and ribosomes (Robb, 1972a; Mims et al., 1975; Hess and Weber, 1976; Harder, 1977).

The cytological events that occur on teliospore germination are not well known. The cytoplasm became less dense and the endoplasmic reticulum (ER) became more extensive (Robb, 1972a; Hess and Weber, 1976). In *Tilletia caries* (DC) Tul. the promycelial cytoplasm resembled that of the teliospore and appeared to undergo little change during promycelial development (Kollmorgen et al., 1979).

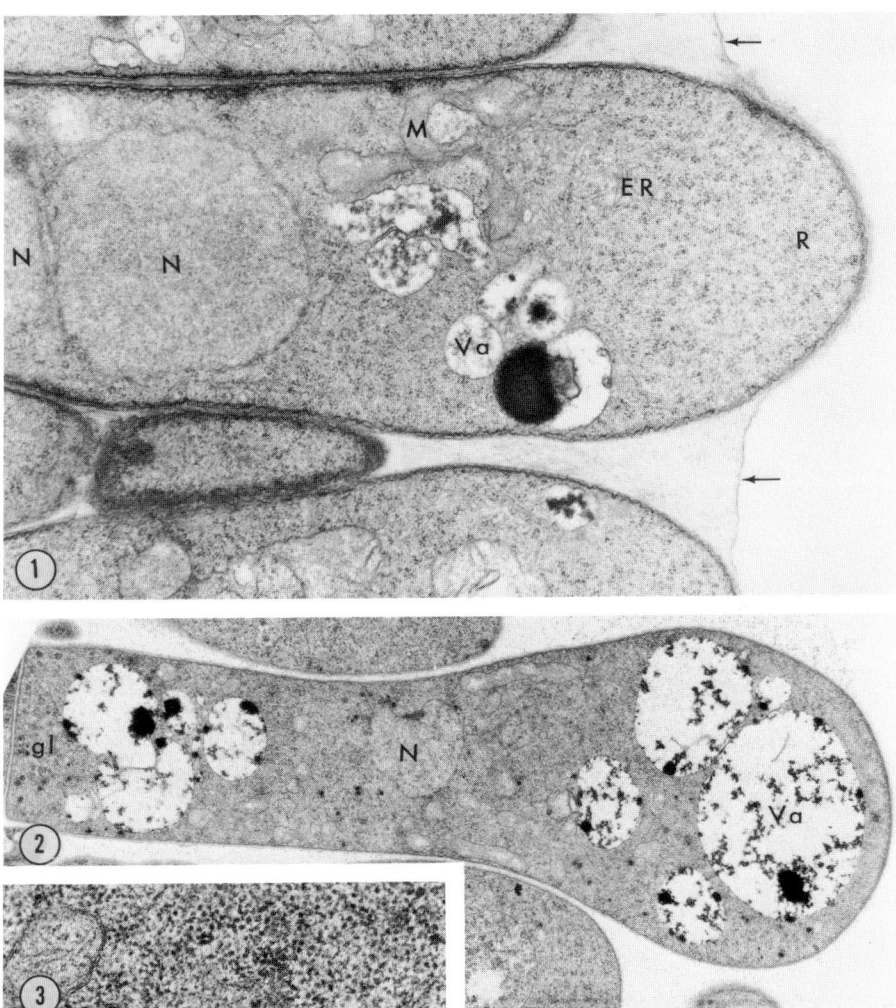

Figs. 2.1–2.3. Longitudinal sections of basidia of *Coprinus cinereus* early and late in the period preceding karyogamy. ER, endoplasmic reticulum; gl, glycogen; M, mitochondrion; N, nucleus; R, ribosomes; Va, vacuole (K. Yoon, unpubl. micrographs). Fig. 2.1 Basidium at −3 days (before maturation) with probable pellicle (arrows) overlying intercellular matrix. The cytoplasm contains few types of organelles ($\times$ 18,100). Fig. 2.2 Basidium at −2 days (4 p.m.) with enlarged apex and apical and basal vacuolation. Numerous free ribosomes and glycogen at the base ($\times$ 7,000). Fig. 2.3 Enlargement of part of Fig. 2.2 showing the dense ribosome concentration ($\times$ 39,500).

However, in *U. maydis* (Ramberg, 1979) the promycelial cytoplasm is considerably more complex than that reported for germinating teliospores of *Ustilago hordei* (Pers.) Lagerh. (Robb, 1972a).

In the late prefusion stage in *C. cinereus* (-2 days; Fig. 2.2) the basidial apex had enlarged, glycogen was evident at the basidial base, and apical and basal vacuoles were present. "Charging" had occurred as the non-vacuolar components of the cytoplasm appeared to have increased since "inception." The most striking feature of this stage was the dense ribosome accumulation (Figs. 2.2, 2.3). The ribosomes were mostly free in the cytoplasm and the ER was not extensive.

"Charging" should probably be interpreted as a phase in which precursors for various synthetic processes enter the basidium, some of which are converted to products visible in the light microscope. This interpretation fits the observation that in many species cells of the gill trama and subhymenium have transparent contents while refractile droplets accumulate within basidia (Corner, 1948). We do not know whether organelles, such as ribosomes, enter the basidium after the basal septum is formed. The early stages of basidial development may be a synthetic period in which products are accumulated within the cell for later phases in basidial differentiation. The accumulation of free ribosomes as observed with the electron microscope (Figs. 2.2, 2.3) and of proteins and ribonucleic acid as demonstrated by light-microscopic cytochemistry (Table 2.1) in *C. cinereus* supports this concept. The limited amount of ER in young basidia suggests that proteins, if formed, are probably accumulated within the cell rather than exported, since the latter process is generally associated with rough ER (Morré and Mollenhauer, 1974).

Storage Product Accumulation

The qualitative distribution of carbohydrates, lipids, protein, and ribonucleic acid in basidia of *C. cinereus* between the prefusion stage and near the time of spore release is shown in Table 2.1. Staining for proteins and ribonucleic acid was intense at the basidial apex beginning with the prefusion stage but declined during basidiospore formation. Carbohydrate granules, presumably glycogen, were present at low levels until late in basidiospore development. Lipid droplets formed in basidia after their inception and were absent from other cells of the hymenium and from the subhymenium. Lipid droplets began to accumulate during the prefusion stage and were abundant within the basidium until late spore formation when they had presumably entered the basidiospores.

Patterns of glycogen and lipid droplet accumulation vary with the species. Lipid droplets accumulated during basidial development and glycogen levels appeared to be low in some species (Wells, 1964a, b; Oláh, 1973). In *Boletus rubinellus* Peck (Beckett et al., 1974, p. 205; and unpublished results) both glycogen and lipid droplets formed in abundance, but in *Schizophyllum commune* Fr. (Wells, 1965; Sundberg, 1977) there were few lipid droplets in the basidia or basidiospores and glycogen was usually absent from the basidium until after

Table 2.1. Distribution of Carbohydrates, Lipids, Protein, and Ribonucleic Acid at Successive Stages in Basidial Development of *Coprinus cinereus*[a].

	Basidial stage				
	Prefusion (−2 days; 10 a.m.[b])	Late prophase I (−1 day, 9 a.m.)	Spore initiation (−1 day, 1 p.m.)	Mid spore formation (−1 day, 4 p.m.)	Late spore formation (−1 day, 10 p.m.)
Carbohydrates[c]	+	+	+	+ to ++	++
Lipids[d]	+	++	++	++	0 to +
Protein[e]	++	++	++	+ to ++	+ to ++
Ribonucleic Acid[f]	++	++	++	+	+

[a]Plastic sections, 2 μm thick (Feder and O'Brien, 1968), were used except for lipids.
[b]6 a.m.–6 p.m. day.
[c]Periodic acid-Shiff reaction ± dimedone blocking (Feder and O'Brien, 1968). Control: periodic acid oxidation omitted.
[d]Sudan IV: acetone-extracted control; fresh, hand sections (Jensen, 1962).
[e]Aniline blue black (Fisher, 1968); pepsin-extracted control for prefusion and spore initiation. Pepsin (1:10,000; Sigma Chemical Co.) 2 mg/ml in 0.02 N HCl, 3 h at 37 °C.
[f]Toluidine blue O (Feder and O'Brien, 1968); ribonuclease (5× recrystalized; Nutritional Biochemical Co.; 0.1 mg/ml in distilled H$_2$O, pH 7.2, 2 h at 55 °C) and hot water controls.

meiosis. Lipid droplets are common in teliospores and their structure has been investigated by Gardner and Hess (1977). Polysaccharides have been localized at the ultrastructural level in mature teliospores of *Puccinia coronata* Cda. *f. sp. avenae* Eriks. (Harder, 1977).

Wall Development in the Basidium

A comparison of wall thickness and reaction to Thiery's silver-protein stain (Table 2.2) in the hymenium and subhymenium of *Coprinus cinereus* at the time of basidiospore formation (McLaughlin, 1974) revealed that the basidial wall appeared to be little differentiated from that of the subhymenium which it resembled in thickness and staining. Walls of the pseudoparaphyses and cystidia were highly differentiated. Their walls stained intensely with silver protein, and those of the pseudoparaphyses were very thin and of the cystidia, very thick. Although the walls of the phragmobasidia of *Auricularia* have been considered undifferentiated, those of the metabasidium of *A. fuscosuccinea* (Mont.) Farl. were thicker than walls of other hymenial and subhymenial cells (McLaughlin, 1980) and may contribute to the ability of *Auricularia* spp. to survive desiccation.

Probasidial wall thickening is characteristic of many species with phragmobasidia, as well as some with holobasidia, and may allow the young basidium to remain dormant for long periods. Teliospores may have complex walls (Allen et al., 1971; Robb, 1972a, Müller et al., 1974; Kohno et al., 1975; Mims et al., 1975; Hess and Weber, 1976; Harder, 1977; Mims, 1977; Mims and Thurston, 1979; Ramberg, 1979) and they have been most extensively studied in the Uredinales (Littlefield and Heath, 1979). Several patterns of formation of teliospore wall ornamentation are now known. Warts of *Puccinia smyrnii* Biv.-Bernh. (Bennell et al., 1978) and spines of *Tranzschelia anemones* (Pers.) Nannf. (Bennell and Henderson, 1978) were deposited within the primary wall adjacent to the plasma membrane. They were subsequently exserted as the outer wall collapsed and a secondary wall was laid down beneath them. The process resembles urediospore formation (Littlefield and Heath, 1979), but elements of the endoplasmic reticulum did not underlie the developing warts of *P. smyrnii* as they did urediospore

Table 2.2. Comparison of Cell Walls in *Coprinus cinereus*.

	Average wall thickness (nm)[a]	Periodic acid–silver-protein reaction[b]
Basidium	85	+
Pseudoparaphysis	35	++ to +++
Cystidium	279	++
Subhymenium and lateral gill trama	80	+ to ++

[a]Each value is the mean of 10 or more measurements.
[b]+, light; ++, medium; +++, heavy (McLaughlin, 1974).

spines. Formation of the spines on teliospores of *Puccinia podophylli* Schw. (Mims and Thurston, 1979) involved cytoplasmic outgrowths at thin spots in the wall with later evacuation of cytoplasm from the spines at least in the tips, while aecioid teliospores resemble aeciospores in the manner of ornament development (Littlefield and Heath, 1979, p. 75).

In double teliospores of the Uredinales (often, but misleadingly, referred to as two-celled teliospores; Donk, 1972) the timing of initiation of wall ornamentation is variable. In *P. smyrnii* (Bennell et al., 1978) the common ornamentation on the double teliospore was initiated prior to septation of the primary teliospore cell to form the two cells of the double teliospore. Thus, the integrated pattern of ornamentation of the double teliospore may be achieved by early wall initiation. In *Puccinia coronata f. sp. avenae* (Harder, 1977) a cap of wall material formed at the terminal end of the primary teliospore cell prior to its division into two cells. But in *T. anemones* (Bennell and Henderson, 1978) septum formation occurred early in the primary teliospore cell and secondary wall development proceeded independently in each cell of the double teliospore. Cells of the double teliospores of *T. anemones* and *Gymnosporangium juniperi-virginianae* Schw. (Mims et al., 1975) separated readily at maturity and the separation process involved a centripetal cleavage through the middle of the septum (Mims et al., 1975).

The clearest understanding of teliospore wall formation has been achieved in the Tilletiales by using several ultrastructural techniques (Gardner et al., 1975; Hess and Weber, 1976). In *Tilletia caries* and *T. controversa* Kühn, three wall layers and a partition layer were laid down in succession (Hess and Weber, 1976). The reticulate ornamentation was deposited within the first formed wall layer at sites where the spore cytoplasm extended into the wall and before the partition and innermost layer were deposited. It is not clear if the process of reticulum formation is different from that reported for spores of the Uredinales (Henderson et al., 1972) but the latter have not been adequately investigated.

The innermost layer of the teliospore wall is usually continuous with that of the metabasidium or promycelium (Kohno et al., 1975; Littlefield and Heath, 1979, p. 87; Ramberg, 1979). In *Tilletia* spp. there was a localized dissolution of the teliospore wall at the site of germ tube emergence (Hess and Weber, 1976), and the germ tube wall was not continuous with the innermost spore wall layer. Rather it appeared to be a new wall layer deposited at the germination site. After germination a major portion of the teliospore wall of *Ustilago maydis* showed a change in texture, suggesting that the wall was being broken down and reutilized by the promycelium (Ramberg, 1979). The promycelial wall was thin in *T. caries* (Kollmorgen et al., 1979) and thick in *U. maydis* (Ramberg, 1979). This difference may be associated with the longevity of the promycelium. In *T. caries* it was short-lived and collapsed after forming basidiospores, while in *U. maydis* it was persistent and produced numerous basidiospores and yeast cells.

A trilaminar layer or pellicle appeared to envelope the gill surface in *C. cinereus* from early gill inception (Fig. 2.1) and was clearly present at later stages in gill development (Figs. 2.6, 2.7). An intercellular matrix was present between the cells (Fig. 2.1). The pellicle was present throughout gill development and enclosed the young basidiospore (Fig. 2.12). The presence of this layer indicates that the

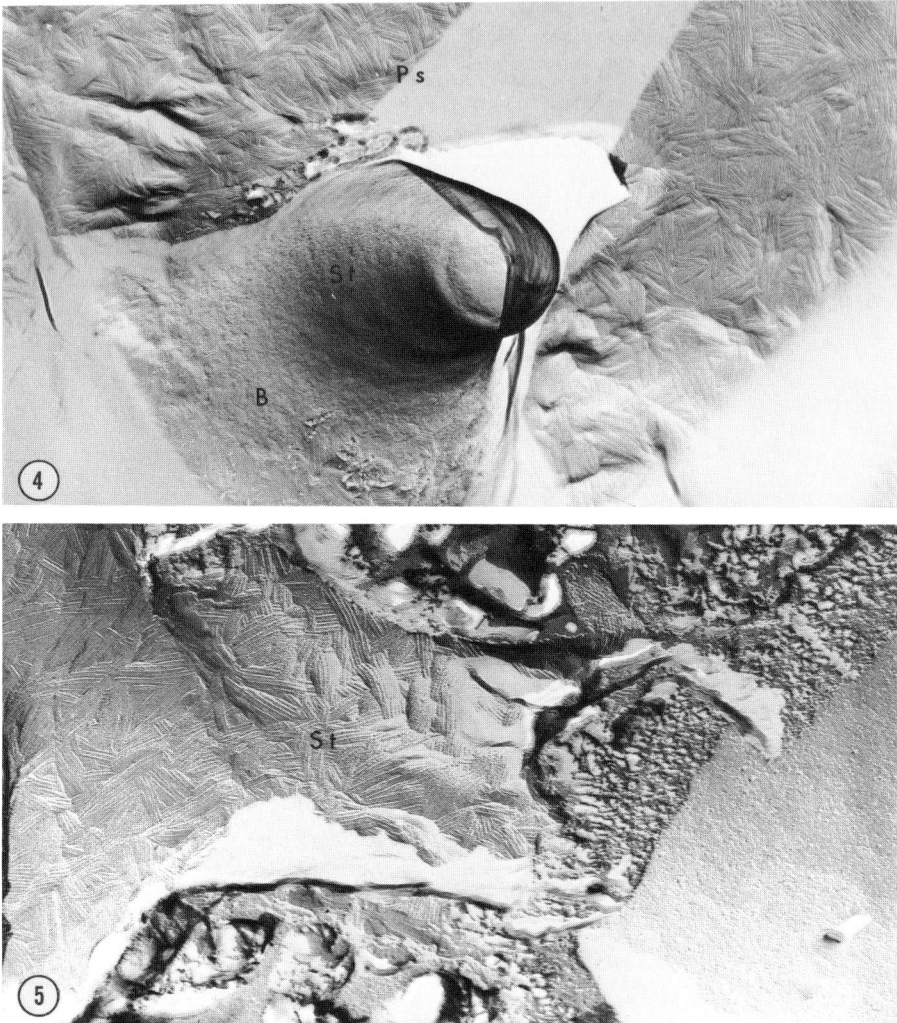

Figs. 2.4–2.5. Frozen-etched preparations of the hymenial surface of *Coprinus cinereus* (A. Seilheimer, unpubl. micrographs). Fig. 2.4 Rodlets cover the surface of the pseudoparaphysis (Ps) and basidium (B) without apparent discontinuity. They are absent from the sterigma (St), which has not begun to form a basidiospore. The pale area on the pseudoparaphysis was caused by blocking of the platinum shadowing by the sterigma; there is a break in the replica adjacent to the sterigma (white area) ($\times$ 33,500). Fig. 2.5 Rodlets cover the basidial apex (left) and the sterigma (St) after basidiospore formation; the fracture plane passes through an inner surface of the basidiospore (right) ($\times$ 62,600).

gill is a tissue. Its cells are not individual hyphal tips, but are united beneath a common membrane, which they form together, and are embedded in a common matrix. A pellicle may enclose basidia of *Boletus rubinellus* (unpubl. results), and a membrane covers the hymenium in *Lentinus edodes* (Berk.) Sing. (Nakai and Ushiyama, 1974). The pellicle and intercellular matrix may provide the medium

Basidial and Basidiospore Development

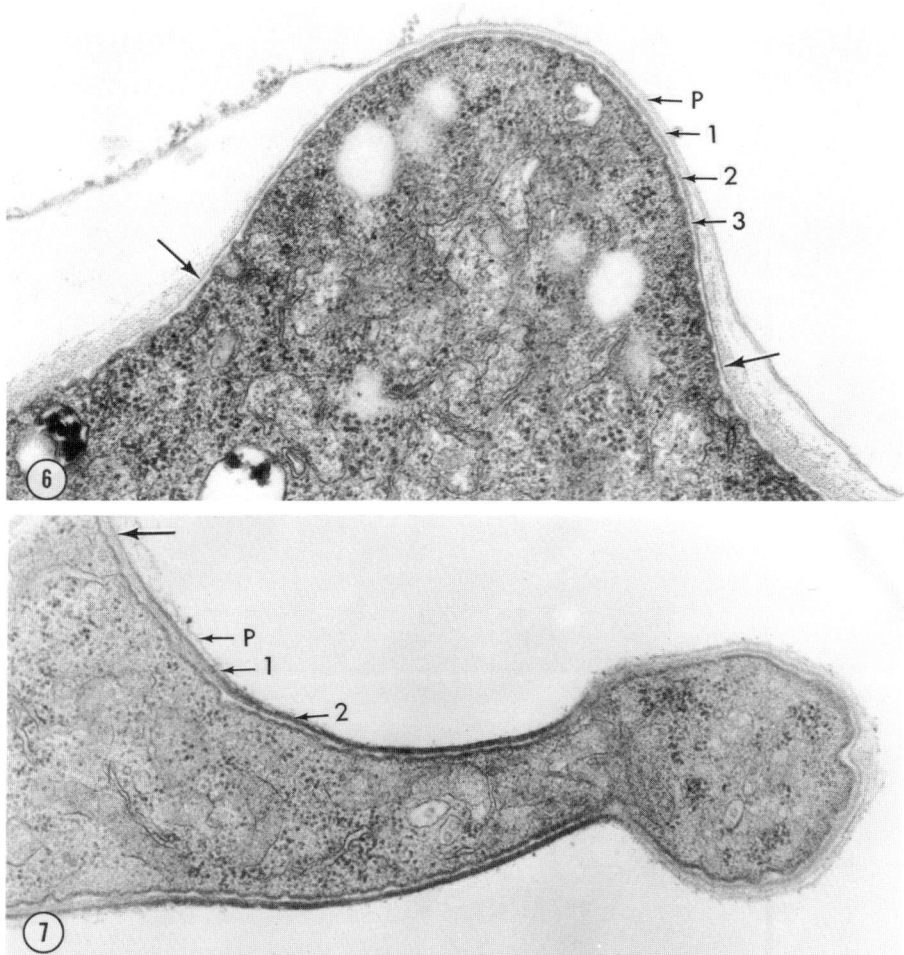

Figs. 2.6–2.7. Longitudinal sections showing sterigmal development in *Coprinus cinereus*. Unless otherwise indicated, these and all subsequent micrographs are unpublished and by the author. Fig. 2.6 The sterigma at initiation has a 3-layered wall (1–3) enclosed by a pellicle (P) which has partly separated from the basidium (left). Wall layer 2, which forms the bulk of the sterigmal wall, terminates in the basidial apex (large arrows) ($\times$ 39,100). Fig. 2.7 The sterigma after basidiospore initiation (right) with its characteristic electron-dense wall (layer 2) which terminates in the basidial apex (large arrow). The pellicle (P) cannot be seen on the sterigma ($\times$ 40,400).

whereby the individual hyphae within the fruitbody are able to communicate with each other to form an organized structure. Since this layer in *C. cinereus* is the product of a group of hyphae rather than a single hypha, it should not be interpreted as a wall layer as was previously done (McLaughlin, 1977a; see wall layer 1).

The surface of the hymenium and sterigmata of *Coprinus cinereus* showed a rodlet pattern in freeze-etch preparations (Figs. 2.4, 2.5). Since the rodlets extended over the surface of the pseudoparaphyses and basidia apparently without discontinuity (Fig. 2.4), it is probable that the rodlets are associated with the pellicle which covers the gill surface; these results seem to agree with those obtained in isolating the rodlet layer from spores of *Trichophyton mentagrophytes* (Robin) Blanchard (Hashimoto et al., 1976). Although the rodlets were evident only at the base of the sterigmata before basidiospore initiation (Fig. 2.4), they were present on the sterigmata (Fig. 2.5) and spore surface (not shown) later. A rodlet layer was reported on the surface of the sterigmata and basidiospores in some, but not all, basidiomycetes examined by Bronchart and Demoulin (1971) and Hess et al. (1972). The rodlet pattern on the basidiospore of *C. cinereus* was weakly developed, which may be related to the tendency of the outermost wall layers to become sticky and fuse together (McLaughlin, 1977a). Water repellence of wild-type conidia of *Neurospora crassa* Shear et Dodge is associated with the presence of rodlets on their surface (Beever and Dempsey, 1978). Rodlets on the hymenium of *C. cinereus* may have a similar function.

The limited investigations of holobasidia and phragmobasidia have revealed a diversity in wall development in hymenia and ornamented teliospores. Further investigations, especially those employing several ultrastructural methods, should lead to a better understanding of these fungi both developmentally and phylogenetically.

Basidial Septation

Two types of septation, primary and secondary, have been recognized within basidia (Donk, 1972). Primary septa are formed in association with nuclear division, while secondary or adventitious septa partition the cytoplasm without an associated nuclear division. However, the relationship between nuclear division and septum formation is not always clear as meiosis may occur within the teliospore in Ustilaginales. The septum at the base of a basidium which delimits it from the subjacent cell is always a primary septum and has the septal pore apparatus characteristic of the species (Fig. 2.8), although it was altered as teliospores matured (Mims et al., 1975; Harder, 1977; Mims, 1977; Mims and Thurston, 1979). Published accounts of primary septa within phragmobasidia, i.e., septa which form during the reduction divisions of meiosis, show an absence of the septal pore apparatus. Wells (1964a, 1964b) demonstrated the absence of septal pores within basidia of *Myxarium nucleatum* Wallr. [= *Exidia nucleata* (Schw.) Burt], and Robb (1972b) reported that the internal septa were complete in *Ustilago hordei*. It has been suggested that primary septa within the basidium always lack the septal pore apparatus (Donk, 1972). However, in *Auricularia fuscosuccinea* each of the primary septa within the basidium (Fig. 2.11) possessed a septal pore apparatus similar to that at the basidial base (McLaughlin, 1979, 1980). Also, serial sectioning of maturing promycelia of *Ustilago maydis* revealed that a minute central pore was present in some septa (Ramberg, 1979).

Khan and Talbot (1976) showed structural differences between primary and

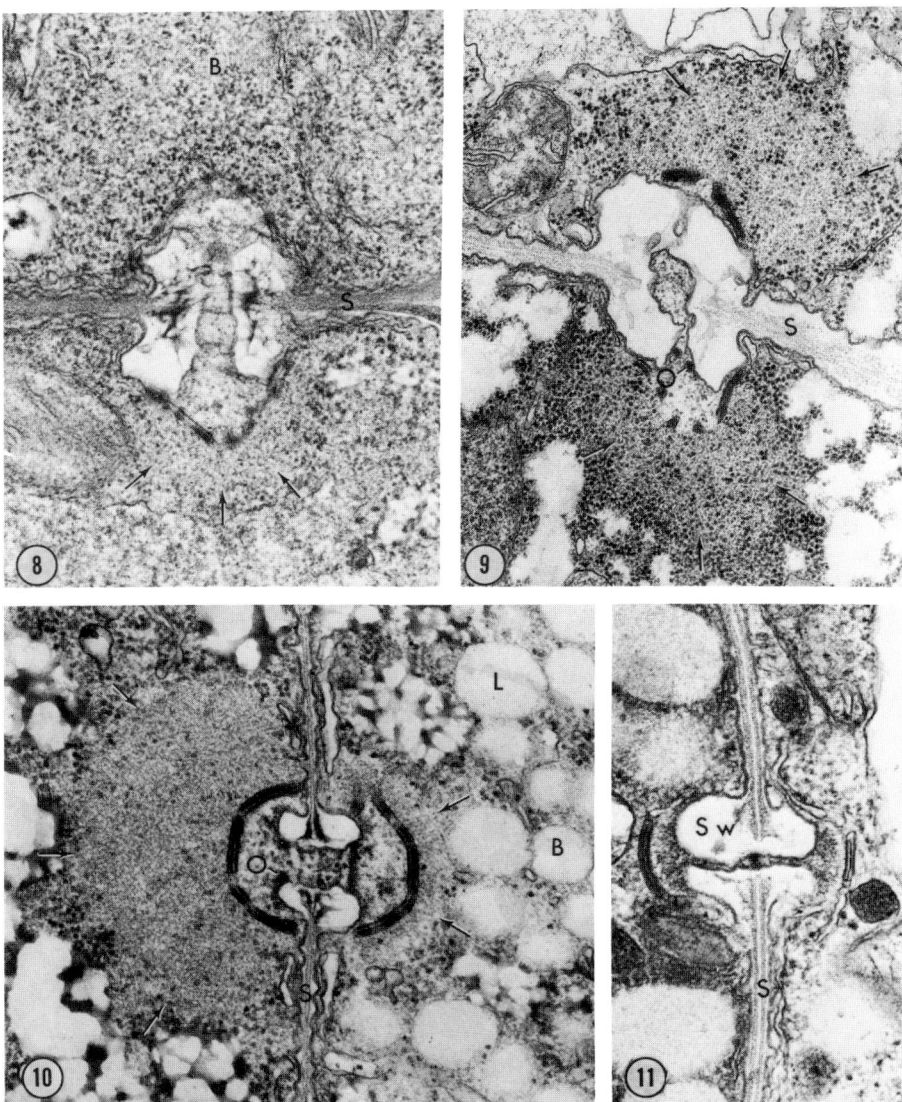

Figs. 2.8–2.11. Longitudinal sections through the septal pore apparatus in the basidium or subhymenium. B, basidium; L, lipid droplet; O, septal pore occlusion; S, septum; Sw, septal swelling. Fig. 2.8 High-voltage electron micrograph of 0.25-μm-thick section of the basidial septum of *Coprinus cinereus* from a gill in which basidiospores were at an early stage in development. An outer cap (arrows) is visible only on the subhymenial side of the septum (× 30,800). Fig. 2.9 Thin section of a subhymenial septum of *C. cinereus* with outer caps (arrows) on both sides of the septum. Pore occlusion present (× 32,300). Fig. 2.10 Septum at the base of a basidium of *Boletus rubinellus* with a small outer cap in the basidium (right; arrows) and a large outer cap (arrows) on the subhymenial side. Pore occlusion present (× 41,400). Fig. 2.11 Septum within a phragmobasidium of *Auricularia fuscosuccinea* with septal pore apparatus like that at the basidial base (× 39,500).

secondary septa in basidia of *Tulasnella* sp. Adventitious septa formed at the base of the protosterigmata and lacked the electron-light central wall layer and septal pore apparatus of primary septa at the basidial base. The adventitious septa are believed to be initiated in a manner different from primary septa.

Basidia of *Myxarium nucleatum* (Wells, 1964b) had a cruciately divided apical region and a vacuolate stalk which was separated by a cross-wall from the apical portion. Wells (1964b) indicated that the cross-wall which delimited the stalk was a primary septum and that a distinction could not be made between primary and secondary septation in this basidium.

The types of septation within basidia of *A. fuscosuccinea, M. nucleatum, Tulasnella* sp., and *U. maydis* are all fundamentally different, i.e., perforate primary septation with a septal pore apparatus in *A. fuscosuccinea*, complete primary septation in *M. nucleatum*, adventitious septa in *Tulasnella* sp., and promycelia with a small pore in some septa in *U. maydis*. Variations in septation may be associated with other differences in basidial development, such as the manner and timing of final vacuolation.

Sterigmal Initiation

The holobasidium attains full size prior to sterigmal initiation (Corner, 1948). Sterigmata begin as broad bumps (Figs. 2.6, 2.27) formed perpendicular to the cell surface typically on the margins of the basidial apex. Subsequently they elongate and curve toward the adaxial side (Figs. 2.16, 2.17). They are typically slender with a characteristic thin, electron-dense wall (Fig. 2.7) (Hoch and Setliff, 1976). The sterigmal wall (20–30 nm thick) in *Coprinus cinereus* was thinner than that of the basidium (Table 2.2) and consisted, at initiation, of perhaps 3 wall layers surrounded by a pellicle (Fig. 2.6). It was not clear whether a pellicle enclosed the sterigma during elongation, but the fact that rodlets covered sterigmata bearing basidiospores (Fig. 2.5) suggested that a pellicle was present later. Wall layer 1 was continuous with the basidial wall and lost as the sterigma elongated (Figs. 2.6, 2.7). Wall layer 2 was electron-dense and was the major component of the sterigmal wall; layer 3 may be an artifact. Layer 2 terminated in the basidial apex (Figs. 2.6, 2.7, arrows). The sterigmal wall is believed to be identical to or continuous with that of the basidium (Hugueney, 1975; Hoch and Setliff, 1976; Perreau, 1976). However, in *C. cinereus* the sterigmal wall is either a modification of the basidial wall or a new wall layer inserted into the basidial apex when sterigmata are initiated. The same results were obtained for sterigmata of *Boletus rubinellus* (Yoon and McLaughlin, unpubl. results).

Sterigmata arose endogenously from the inner part of the basidial wall and, thus, broke through its outer layer, in *Thanatephorus cucumeris* (Frank) Donk (Tu et al., 1977), *Auricularia fuscosuccinea* (McLaughlin, 1979), *Tulasnella* sp. (McLaughlin, 1979), and *Ustilago maydis* (Ramberg, 1979). In the latter species the sterigmata were short and surrounded by a collar derived from the outer layer of the promycelial wall. In *C. cinereus* and *B. rubinellus* no rupture was observed in the outer layer of the basidial wall as the sterigmata were initiated.

Cytoplasmic events in sterigmal initiation are still unclear (Fig. 2.6). Occasional small vesicles (30–50 nm in diameter) have been observed in serial sections of sterigmata of *C. cinereus* early in initiation, but there was no evidence of microtubule involvement in the initiation process. At later stages in development when sterigmata had elongated, microtubules extended throughout their length, and vesicles, presumably derived from Golgi cisternae, appeared to contribute to apical growth of their walls (McLaughlin, 1973; Hoch and Setliff, 1976). The sterigmal apex at this stage had few or no ribosomes or other organelles; this apical stratification is reminiscent of hyphal tip growth (McLaughlin, 1973; Hoch and Setliff, 1976).

Basidiospore Formation

The development of the basidiospore involves a complex interaction of cytoplasm and wall, a subject which has been recently treated elsewhere (McLaughlin, 1977a, Sundberg, 1977). The wall is typically multilayered and a complex terminology exists for these layers (Perreau-Bertrand, 1967; Kühner, 1973; Clemençon, 1975a). There is disagreement among different authors on the ontogeny of the layers (Perreau, 1976; McLaughlin, 1977a). The literature on basidiospore walls is extensive and confusing, and many variations in wall structure have been reported in hymenomycetes and gasteromycetes (e.g., Perreau-Bertrand, 1967; Kühner, 1973; Clemençon, 1974, 1977; Keller, 1974, 1977; Bronchart et al., 1975). In much of the earlier work mature spores were fixed with permanganate, and the results were often poor, or important details may have been lost through oxidation by the fixative. An understanding of basidiospore morphogenesis will require careful developmental studies, and current methods of fixation with aldehyde and osmium may need to be combined with other methods such as freeze-etching and freeze-substitution (Howard and Aist, 1979) to obtain optimal results on cytoplasm and wall. Dormant spores have been successfully fixed with aldehydes by cracking the wall during fixation (Greuter and Rast, 1975).

The typical basidiospore is asymmetrically positioned on its sterigma. In the genesis of this asymmetry, the spore, which begins as a spherical enlargement of the sterigmal apex (Fig. 2.12), grows toward the abaxial side (Fig. 2.13). In *C. cinereus* it then enlarged symmetrically (Fig. 2.15) and finally elongated (McLaughlin, 1977a). Changes in spore shape were related to changes in spore wall layers. Development of the symmetrical spore (Fig. 2.15) was associated with the formation of the outermost, electron-dense wall layer which broke or dissolved at the spore apex as the spore began to elongate. Variation in basidiospore form has been attributed to acropetal hardening and thickening of the wall with different rates of wall setting occurring in different species (Corner, 1948). While wall hardening may account for the final form of the spore, a distinction should be drawn between transient and permanent wall setting. Transient wall setting may account for distinctive spore morphology at early stages in development, while permanent wall setting may not begin until late in sporogenesis (McLaughlin, 1977a). Thickening of the wall is not synonymous with wall setting because thin-

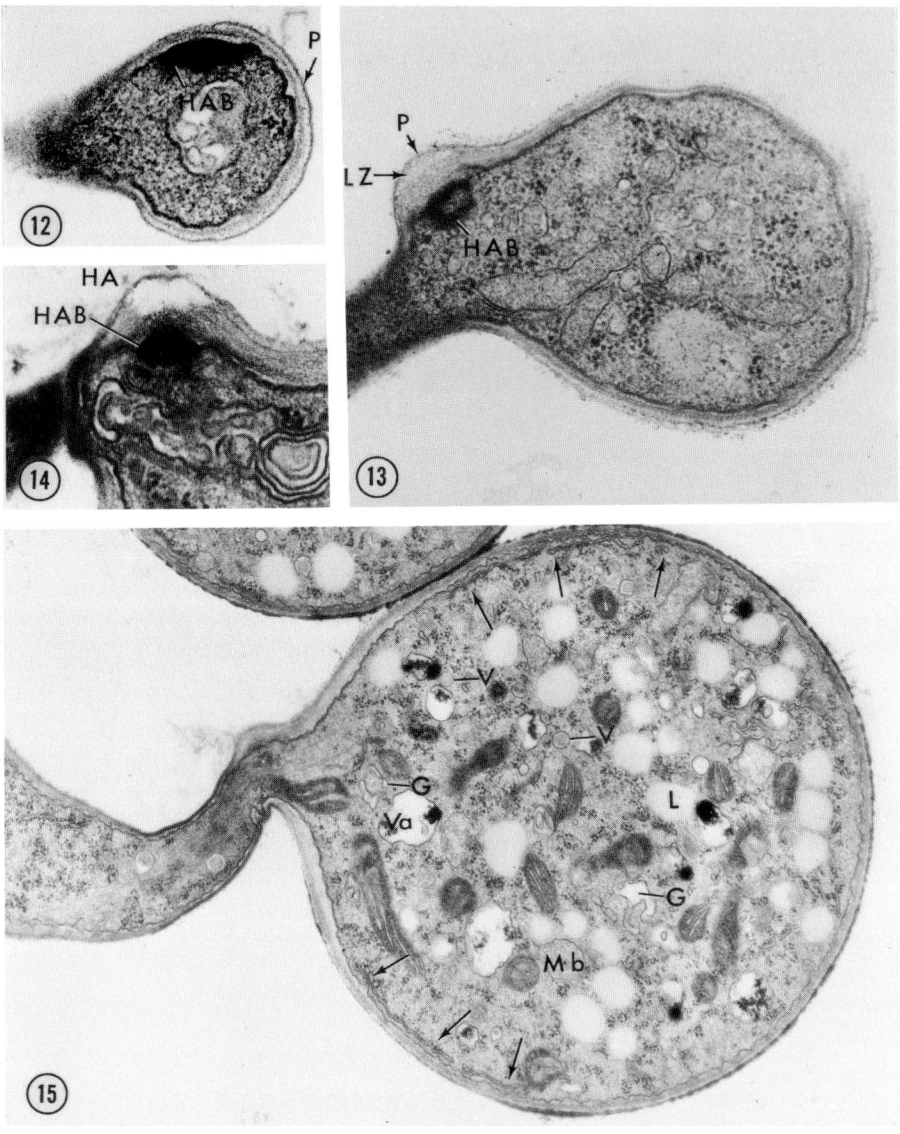

Figs. 2.12–2.15. Longitudinal sections of 3 successive stages in basidiospore initiation in *Coprinus cinereus* with sterigmata to the left and adaxial surface on the upper side of each spore. G, probable Golgi cisterna; HA, hilar appendix; HAB, hilar appendix body; L, lipid droplet; LZ, electron light zone; Mb, microbody; P, pellicle; V, vesicle; Va, vacuole. Figs. 2.13, 2.15 From McLaughlin (1977a). Fig. 2.12 Stage 1: inception. Apex of the sterigma just beginning to enlarge. Hilar appendix body appressed to wall at future site of the electron-light zone of the hilar appendix (× 39,100). Fig. 2.13 Stage 2: asymmetric growth. An electron-light zone is differentiating within the hilar appendix wall adjacent to the hilar appendix body (× 32,300). Fig. 2.14 Same basidiospore as Fig. 2.15 but a median section through the fully developed electron-light zone in the hilar appendix wall and the now cone-shaped hilar appendix body (× 39,500). Fig. 2.15 Stage 3: equal enlargement. The basidiospore enlarges equally in length and width and a peripheral layer of rough endoplasmic reticulum (arrows) lines the lower spore wall during this process. Note the increased complexity of the cytoplasm compared with the earlier stages (× 19,200).

ner areas of the wall may have distinctive shapes at early and late stages in ontogeny (McLaughlin, 1977a), and a correlation between wall thickness and spore asymmetry may be lacking (Burge, 1979).

The basis for spore morphology is to be sought in an understanding of the relationship between wall structure and chemistry. The first study to integrate ultrastructure and chemistry of the basidiospore wall is that of Rast and Hollenstein (1977). They demonstrated a 3-layered wall in *Agaricus bisporus* (Lange) Imbach, the outer layer of which was responsible for cell shape and was impregnated with melanin, which has also been demonstrated in a second species of *Agaricus* (Garcia Mendoza et al., 1979). The underlying layer contained chitin fibrils embedded in a β-glucan and protein matrix and resembled the hyphal wall. A combination of ontogenetic studies with the approach of Rast and Hollenstein (1977) may unravel the terminological problems mentioned above.

Basidiospore wall development is of phylogenetic interest. The basidium is considered to be derived from the ascus by externalization of meiospore production (Kühner, 1973). Some authors have considered the basidiospore to be internally formed like the ascospore, with the outermost basidiospore wall layers derived from the basidial wall and additional wall layers added centripetally (see McLaughlin, 1977a). In fact, some of the external layers of the spore wall may be deposited between the pellicle and the initial wall layers which are continuous with the sterigmal wall (Perreau, 1976; McLaughlin, 1977a). However, as indicated above, there is considerable variation in the manner of basidiospore wall formation in Homobasidiomycetes. In *Coprinus cinereus* it is not clear whether the sterigmal wall should be considered a continuation of the basidial wall. Ascospore walls are formed in an ascus vesicle which surrounds the young spore, and in which the wall is deposited (Beckett et al., 1974). Two lines of evidence could relate the origin of the basidiospore to the ascospore: the origin of the ascus vesicle and the manner in which wall layers are deposited. In *Taphrina deformans* (Berk.) Tul., the genus which has been suggested to be the closest living relative of the ancestral protobasidiomycete (Savile, 1968), an ascus vesicle is not formed; rather, the ascospore-delimiting membrane arises from the plasma membrane (Syrop and Beckett, 1972). Externalization of spore production in this simple ascomycete, which has been reported to occur under abnormal conditions (Savile, 1968), could lead to the basidiospore. This step might readily occur in species of *Taphrina*, since the ascospore-delimiting membrane is in continuity with the plasma membrane. In advanced ascomycetes the substructure of the ascospore-delimiting membrane resembles the ascus plasma membrane (Bracker, 1967), the ascus vesicle is found initially near the plasma membrane (Ashton and Moens, 1979), and some ascomycetes show continuity of these membranes (Campbell, 1973), which suggests a related, even if not identical, origin of the ascus vesicle to the ascospore-delimiting membrane in *T. deformans*. The manner of deposition of the ascospore wall may vary (Campbell, 1973). Its significance may be a moot point with respect to the origin of the basidiospore wall. If a primitive ascomycete were to give rise to its spores exogenously, all transport to the spore wall would then be through the spore plasm even if some of the secretory products would normally arise in the epiplasm. That the origin of the ascus vesicle in Euascomycetes may be independent of the spindle pole body, unlike the situation in hemiascomycetous yeasts (Ashton and Moens, 1979), may also have significance for basidiospore evolution. When basidiospores first evolved, if they continued to be surrounded externally by the basidial wall, the spore wall would have been laid down within an extension of the basidial wall and the spore would still be, in a sense, endogenous although it no longer resided within the main body of the basidium. A hypothesis on the development of true exospores in Agaricales has been presented by Clemençon (1977).

The spherical spore primordium, or apophysis, develops into the hilar appendix or apiculus. There is no agreement as to which of these two terms is preferable (e.g., Hugueney, 1975; Perreau, 1977). Hilar appendix is used here following the definition of Pegler and Young (1971, 1979), but it should be noted that the hilar appendix on mature, hymenomycetous spores is derived from only a part of the apophysis, the other part having developed into the spore. Thus, the hilar appendix is a modified remnant of the apophysis, which bears the site of formation of the droplet associated with basidiospore discharge (Hugueney, 1975; Yoon and McLaughlin, 1979). In *Russula* spp. the hilar appendix grew larger than the apophysis and became cylindrical (Burge, 1979).

In *C. cinereus* an electron-dense region, the hilar appendix body, formed on the adaxial side of the primordium at the time of its initiation (Fig. 2.12) (McLaughlin, 1977a). As the spore grew asymmetrically toward its abaxial side, the hilar appendix body rested on a wall thickening and an electron-light zone was initiated adjacent to it within the wall (Fig. 2.13). At the stage of equal enlargement (Fig. 2.15) this zone appeared fully formed as an electron-light compartment in the wall of the hilar appendix (Fig. 2.14). In *Coprinus* the hilar appendix body disappeared as the spore reached full size (Hugueney, 1975), but it was absent at an earlier stage in *Boletus rubinellus* spores (Yoon and McLaughlin, 1979). It may be involved in synthesis of the discharge apparatus, and has been reported only from several *Coprinus* spp. (Hugueney, 1975; McLaughlin, 1977a) and *B. rubinellus* (Yoon and McLaughlin, 1979).

The electron-light compartment in the adaxial wall of the hilar appendix of *Coprinus* spp. and *B. rubinellus* has not been observed in thin-walled basidiospores (Wells, 1965; Nakai and Ushiyama, 1974). In the thick-walled spores of *B. rubinellus* the droplet formed before separation of the spore cytoplasm from that of the sterigma (Yoon and McLaughlin, 1979), but apparently developed later in thin-walled spores (Wells, 1965).

The basidiospore cytoplasm is first separated from that of the sterigma by a centripetal invagination of the plasma membrane (McLaughlin, unpubl. results), which in some spores developed into an electron-light zone (Wells, 1965; Nakai and Ushiyama, 1974; Hoch and Setliff, 1976). Subsequent differentiation at the hilum apparently depends on the type of basidiospore. *Schizophyllum commune* (Wells, 1965) and *Lentinus edodes* (Nakai and Ushiyama, 1974) with thin-walled spores formed an electron-light, parahilar area adjacent to the hilum at the spore base. Species of *Coprinus* and *Boletus rubinellus* with thick-walled spores (Oláh and Reisinger, 1974; Hugueney, 1975; Yoon and McLaughlin, 1979) formed two plugs, a hilar plug at the spore base and a sterigmal plug. There is considerable variation in the structure of the plug in the hilar appendix of hymenomycetes (Kühner, 1973; Keller, 1974; Clemençon, 1977). Two types of hilum morphology have been reported after spore discharge. A nodulose type was found in hyaline spores and an open pore type in spores with pigmented walls (Pegler and Young, 1969, 1971). The drop which forms at the hilar appendix before discharge survived fixation (Corner, 1948) and sectioning (Wells, 1965; Oláh and Reisinger, 1974; Yoon and McLaughlin, 1979) and appeared to contain polysaccharides (Oláh and Reisinger, 1974). These results support Wells' (1965) hypothesis that the drop may be an area capable of exerting high osmotic pressure and swelling, because of water uptake, just before discharge. After discharge the hilar appendix has a basal scar, the hilum, and an adaxial scar, the *punctum lacrymans* (Latin: weeping point) (Hugueney, 1972, 1975) where the droplet was released.

Dissociation of spores from the sterigmata occurs differently in gasteromycetes than in hymenomycetes. In hymenomycetes abscission occurs through the septum which separated the sterigma from the spore, while in the gasteromycetes the break occurs below this septum in the sterigmal wall (Perreau, 1977). Thus, gasteromycetous spores may be released with part of the sterigmal wall still attached. This wall has been incorrectly referred to as a hilar appendage, but it is a sterigmal appendage (Pegler and Young, 1979) or a peduncle (Perreau, 1977). Basidiospore wall differentiation may extend part way down the sterigma in some gasteromycetes (see Perreau, 1977) and the hilar or endosporic plug is not present in all groups. In members of the Phallales examined by Perreau (1977) the hilar plug was absent and the basidiospores resembled an endospore in the manner of inner wall formation.

Cytoplasmic Differentiation at the Septal Pore

Differentiation of the cytoplasm has been reported both within the septal pore and around the septal pore caps at the base of the basidium. An outer cap or cytoplasmic zone surrounding the septal pore cap was first illustrated in basidia of *Agaricus bisporus* (Thielke, 1972). It has since been found in numerous hymenomycetes but often only on the subhymenial side of the basidial septum as in *C. cinereus* (Fig. 2.8), whose septum is shown here in a high-voltage micrograph. It was well-developed around the septal caps on both sides of subhymenial septa (Fig. 2.9) (McLaughlin, 1974). The outer cap was rarely seen in basidia of *C. cinereus* at the time of spore initiation, using standard thin sections. In *Boletus rubinellus* outer caps were present around the septal caps on both the basidial and subhymenial side of the septum but the outer cap was smaller on the basidial side (Fig. 2.10). The outer cap was not observed on the basidial side of the septum in *A. bisporus* by Craig et al. (1977) in contrast to the findings of Thielke (1972, 1978), which suggests that the differentiation of the outer cap may depend on the stage of basidial development. It was present only on the subhymenial side of hymenial/subhymenial septa in some species (Gull, 1976; Wells, 1978; Khan and

Kimbrough, 1979). The outer cap was sometimes enclosed by ER, and it consisted of fibrils (McLaughlin, 1974; Gull, 1976; Khan and Kimbrough, 1979) or a region often densely packed with ribosomes (Thielke, 1972; Craig et al., 1977) or was granular (Wells, 1978). It is possible that filaments were obscured in outer caps with many ribosomes. In *C. cinereus* the outer caps stained for ribonucleic acids and protein, but not for carbohydrates (McLaughlin, 1974). The function of the outer cap is unclear. It has been suggested that it isolates meiotic cells from the rest of the fruitbody (Gull, 1976; Craig et al., 1977), that it controls organelle movement (Khan and Kimbrough, 1979), or that it is involved with active transport through the septal pore (McLaughlin, 1974). The outer cap seems to be associated with fruitbodies. It was found within generative hyphae in the interior of *Pterula* sp. fruitbodies (McLaughlin and McLaughlin, 1977), and it is unlikely that these internal hyphae are involved with basidial formation. I suggest that all of the outer caps contain microfilaments and may have a role in active transport within the fruitbody.

Membranes or pore rings and electron-dense occlusions (Flegler et al., 1976) occur in the septal pore at the base of the basidium (Thielke 1972, 1978; Wells, 1978; Khan and Kimbrough, 1979) (Fig. 2.10). Membranes, pore occlusions and pore rings in conjunction with microfilaments have been suggested to regulate cytoplasmic streaming (Flegler et al., 1976; Wells, 1978; Khan and Kimbrough, 1979). The septal pore caps may be different in the fruitbody than in the mycelium. *Dacrymyces stillatus* Nees ex Fr. had septal pore caps with a small central pore in the fruitbody and imperforate septal caps in the mycelium (Flegler et al., 1976). A similar pore occurred in the septal pore cap at the basal septum of the basidium of *Auricularia fuscosuccinea* (McLaughlin, 1979).

After spore discharge the basidium collapses, the septal pore apparatus disintegrates and the pore seals (Wells, 1964a; Nakai and Ushiyama, 1974). In *Pholiota terrestris* Overholts the sealing of the septal pore involved expansion of the septal swellings (Wells, 1978). A secondary wall may form over the septal pore in the subbasidial cell (Wells, 1964a; Khan and Talbot, 1976).

Vacuolation of Maturing Basidia

As basidiospores form, a large vacuole, usually at the base of the basidium, gradually expands to fill the cell. The vacuolation process occurs in holobasidia (Corner, 1948; Wells, 1965; Hoch and Setliff, 1976) and phragmobasidia (Wells, 1964b). Dense material of unknown nature and function may be present within the vacuoles (Thielke, 1967, 1969) (Fig. 2.16), or they may contain granular or fibrillar material (Thielke, 1978; Ramberg, 1979).

A careful analysis of the relationship between basidiospore and basidial size and the degree of vacuolation at the time of spore initiation led Corner (1948) to hypothesize that basidiospore formation and size depended on the expulsion of the basidial protoplasm into the spore by the enlarging vacuole. He termed this process the "ampoule effect," with the basidium acting as the ampoule and the basal vacuole as the driving force to fill the spores. There are a number of problems with this concept. First, it tends to view basidiospore growth as a passive process; the need for synthesis and energy in construction of the basidiospore are de-emphasized. Second, cell components enter the basidiospore in a definite sequence (McLaughlin, 1977a) and are not randomly pushed into it. Nuclei move actively and their movements appear to be under the control of the spindle pole body or nucleus-associated organelle (Girbardt, 1968; Girbardt and Hädrich, 1975). Third, Corner (1948) described several exceptional cases in which the basidia vacuolated apically and only a part of the cytoplasm entered the spores. These cases suggest that basidiospore size is not controlled by the volume of protoplasm in the basidium less the vacuolar volume. Microfilaments have been implicated in the control of cytoplasmic movement (Hepler and Palevitz, 1974) and have been reported in basidia, where they may have a similar function (Gull, 1975).

The "ampoule effect" may indicate that turgor pressure is necessary for basidial growth and that there has been selection for a relationship between basidiospore size and the volume of basidial protoplasm, less the vacuolar volume. The role of turgor pressure in hyphal growth is still unproven (Bartnicki-Garcia, 1973), but it may provide a force for hyphal extension or may be needed to keep the plasma membrane in contact with the wall to allow synthetic processes to occur. The relationship between the volume of basidial protoplasm, less the vacuolar volume, and spore size may be best explained by natural selection in which the basidial cytoplasm is most efficiently utilized when most of it is packaged within the spore.

Vacuolation of phragmobasidia differs from that in holobasidia and the pattern of vacuolation may be related to the ecological adaptations of the organism. In *Auricularia fuscosuccinea* (McLaughlin, 1979, 1980) the basidial compartments were evacuated in basipetal sequence. A large, postmeiotic increase in cytoplasmic volume accompanied sterigmal development on each compartment. Vacuolar enlargement began at the base of a compartment only when sterigmal formation was well advanced. In *Puccinia malvacearum* (McLaughlin and O'Donnell, unpubl. results) vacuolation of the basidial compartments began early in sterigmal development and each compartment matured independently of the others. In *Ustilago maydis* (Ramberg, 1979) evacuation of a basidial compartment was not completed during formation of the first basidiospore. Incomplete vacuolation appears to be related to the capacity of this basidium to form additional spores. Whether these differences in final vacuolation of phragmobasidia are characteristic of each order has yet to be determined.

Endomembrane System

The endoplasmic reticulum (ER) and Golgi apparatus are part of the endomembrane system which also includes the nuclear and plasma membrane, various types of vesicles, vacuoles, and lysosomes. This membrane system operates along two major routes within the cell, the first route being associated with secretion at the plasma membrane and the second, with vacuole formation (Morré and Mollenhauer, 1974). The role of the endomembrane system in basidial development is incompletely known.

Various ER configurations have been reported in basidia. Perinuclear cisternae of the ER occurred in *Agaricus bisporus* during meiosis and may be involved in reformation of the nuclear membrane (Thielke, 1968a, 1968b). In *Coprinus radiatus* (Bolt. ex Fr.) S.F. Gray the perinuclear cisternae associated with dividing nuclei were connected to membrane coils which have been suggested to result from a reduction in nuclear volume during meiosis (Thielke, 1968b, 1974; Lerbs, 1971). Perinuclear cisternae surrounded postmeiotic nuclei in many Agaricales (Clemençon, 1969) and in *Poria latemarginata* (Dur. et Mont.) Cke. (Setliff et al., 1974). The ER may exist as sheets of parallel lamellae, and, in this form, it has been observed in basidia from prefusion to postmeiotic stages in *Myxarium nucleatum* (Wells, 1964a, 1964b) and *Boletus rubinellus* (Beckett et al., 1974, p. 205). It is a form of rough ER (McLaughlin, 1970 and unpubl. results). The ER has been reported to line the basidial wall in *C. radiatus* (Lerbs, 1971; Thielke, 1974) and *B. rubinellus* (Beckett et al., 1974, p. 205). The functions of these various ER configurations are uncertain. A band of ER lined the lower part of the basidiospore wall of *C. cinereus* (Fig. 2.15), but only when the spore was enlarging equally in length and width (McLaughlin, 1977a). It may contribute to cell expansion through protein secretion. In promycelia of *Ustilago maydis* rough ER was frequently associated with the plasma membrane and septa (Ramberg, 1979), and had an asymmetric distribution of ribosomes like the ER which lined the lower spore wall in *C. cinereus*. The rough ER in *U. maydis* may also have a secretory function. In developing teliospores of *Puccinia smyrnii* ER lined the wall at sites where ornamentation was not formed (Bennell et al., 1978).

The elaboration of the endomembrane system late in basidial development in *Coprinus cinereus* has been extensively studied (McLaughlin, 1972, 1973, 1974, 1977a, 1977b). Clemençon (1969) reported that proliferation of the ER in basidia of the Agaricales was initiated at karyogamy. Changes in the ER and the first appearance of Golgi cisternae have been investigated early and late in first meiotic prophase in *C. cinereus*. There appeared to be some increase in rough ER toward the end of prophase, but Golgi cisternae were not observed with certainty until sterigmal initiation. In the postmeiotic basidium rough ER was attached to the nuclear membrane and extended throughout the cell (Figs. 2.16, 2.24), and areas of probable Golgi activity were common. The failure of some authors to find Golgi cisternae may be because the cisternae are not present or well-developed at all stages in basidial development.

ER is usually poorly developed in ungerminated basidiospores. However, in *Fomes fomentarius* (Fr.) Kickx., ER cisternae were common in ungerminated spores but rare in germinated ones (Tsuneda and Kennedy, 1978).

The Golgi apparatus in most fungi consists of individual cisternae which are not stacked to form dictyosomes. Golgi cisternae are not easy to recognize and have been reported in few basidia (McLaughlin 1972, 1973, 1974; Hoch and Setliff, 1976). The different forms of Golgi cisternae in *C. cinereus* have been determined using serial thin and thick sections for standard and high-voltage electron microscopy, respectively. This approach was combined with an attempt to

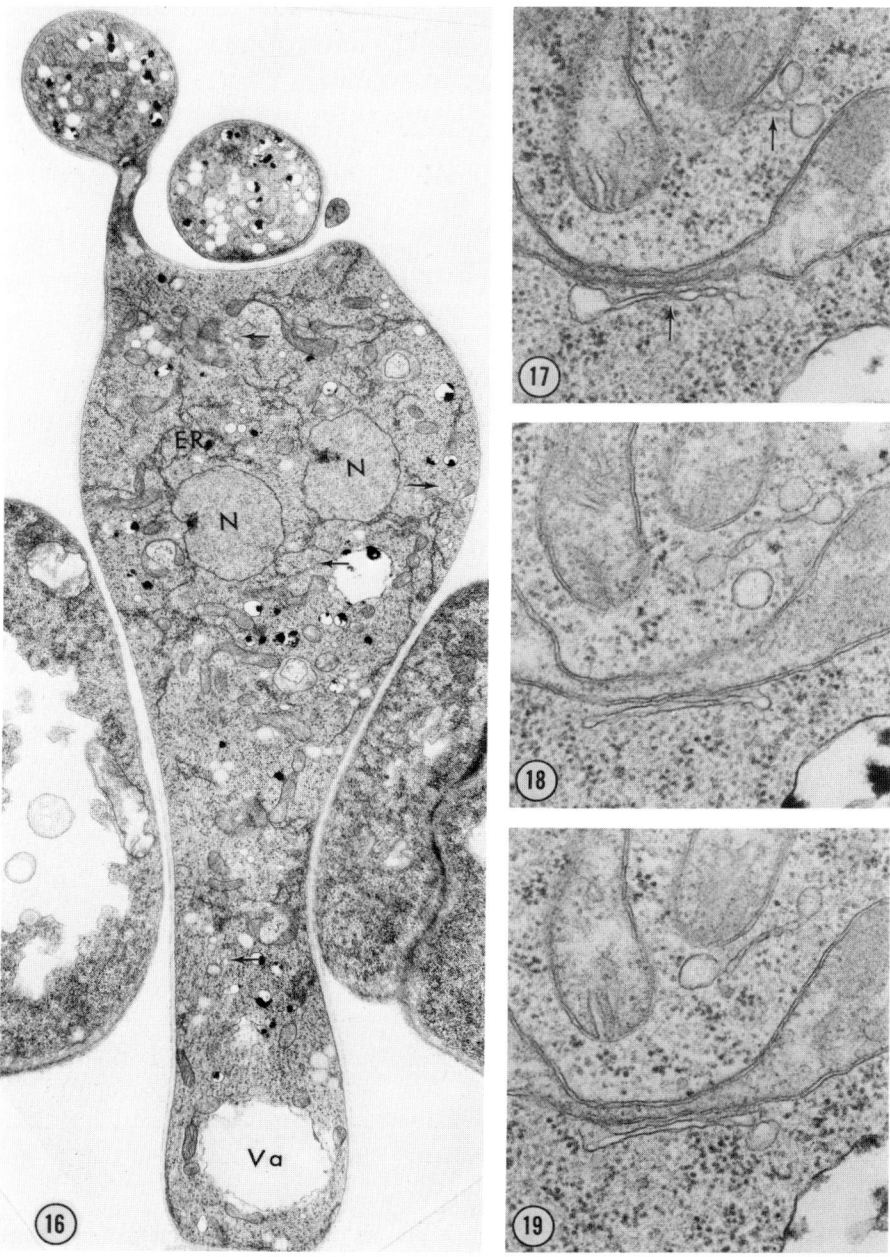

Figs. 2.16–2.19. Distribution of the endoplasmic reticulum and areas of probable Golgi apparatus activity (arrows) in a basidium of *Coprinus cinereus*, in longitudinal section, with a stage-3 spore. Golgi activity is indicated by cisternae or clusters of vesicles. ER, endoplasmic reticulum; N, nucleus; Va, vacuole ($\times$ 5,400). Figs. 2.17–2.19. Three serial, thin sections of two Golgi cisternae (arrows) illustrating the plate form of the cisternae in the postmeiotic basidium of *Coprinus cinereus* ($\times$ 40,400).

determine the initial form of the Golgi cisternae with prolonged treatment with osmium, and with silver-protein staining for carbohydrates to identify products contained in Golgi cisternae and vesicles (McLaughlin, 1974). This research seeks to discover the origin of Golgi cisternae, the transformations they undergo, and the roles they play in the basidium.

Three forms of Golgi cisternae have been identified: "ring," plate and tubular-

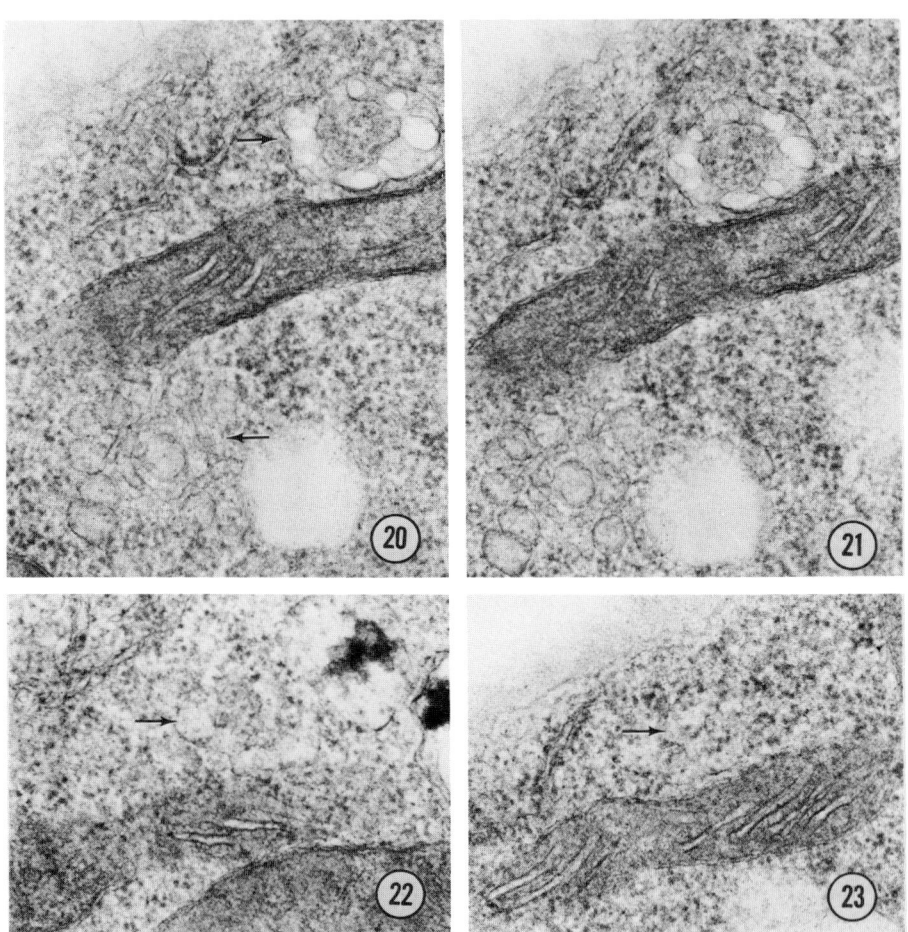

Figs. 2.20–2.23. High-voltage micrographs of 0.25-μm-thick sections of the Golgi cisternae of *Coprinus cinereus*. Figs. 2.20, 2.21 Stereo pair of the "ring" or spherical form of the Golgi cisternae (above, arrow) and of the tubular-vesicular form (below, arrow). A stereoviewer should be used to see the 3-dimensional form of the cisternae ($\times$ 46,000). Figs. 2.22, 2.23 Serial sections which precede (Fig. 2.22) and follow (Fig. 2.23) the section of the "ring" cisterna shown in Fig. 2.20. Because of the thickness of the sections ribosomes appear to be present on the cisterna (arrows), but they are not attached to it when seen in median section (Fig. 2.20) ($\times$ 46,000).

vesicular. "Ring" cisternae (Figs. 2.20–2.23) were common in basidia. Serial sections showed that this form was essentially a fenestrated sphere. The plate cisternae (Figs. 2.17–2.19) may be nearly flattened to cup-shaped with small and large vesicles attached and some fenestrations. Stereoscopic views of the tubular-vesicular cisternae (Figs. 2.20, 2.21) showed a branched, tubular structure with attached and free vesicles. I suggest that the "ring" cisternae is the immature form, and that it gives rise to the plate form and then becomes dispersed into tubules and vesicles. These transformations are similar to those that occur in cisternae within dictyosomes as they pass from the forming to the secretory face (Morré and Mollenhauer, 1974).

Long incubation in osmium at 40 °C has been used to stain the forming face of the Golgi dictyosome and sometimes adjacent ER in animal and plant cells (Friend and Murray, 1965; Friend, 1969; Morré and Mollenhauer, 1974). This method produced electron-dense deposits in the ER and nuclear membrane in *Coprinus cinereus* basidia (Fig. 2.24), but cisternae which resembled Golgi cisternae in size showed little or no stain (Fig. 2.25). The morphology of the initial form of the cisternae is uncertain.

Ultrastructural localization of carbohydrates in *C. cinereus* by the silver-protein technique has shown that compounds, possibly polysaccharides and mucopolysaccharides, accumulate in Golgi vesicles on the cisternae (Fig. 2.26) and are transported to the basidiospore wall (McLaughlin, 1974). The basidial wall was essentially unstained by this method and, therefore, has a different composition from that of the spore.

Two sizes of vesicles are present in the developing basidiospore, small (about 50–60 nm diameter) and large (about 80–100 nm diameter) (McLaughlin, 1974, 1977a; Hoch and Setliff, 1976). The larger vesicles resemble those reported in hyphal regeneration from protoplasts of *Schizophyllum commune*, and are perhaps associated with formation of a glucan layer in the wall (van der Valk and Wessels, 1976). Small vesicles or those present within multivesicular bodies may be chitosomes, which may be involved with the transport of chitin synthetase to the wall (Bracker et al., 1976). Small vesicles and multivesicular bodies have been reported in other basidia (Oláh et al., 1977).

Other components of the endomembrane system that have been reported in basidia include ER vesicles and siderophilous (carminophilous) granules. The ER vesicles were formed near the start of meiosis (Clemençon, 1969), and they showed a great variation in number and size. Some of these vesicles may be microbodies and others, vacuoles or lysosomes as originally suggested. The siderophilous granules were found in several genera of hymenomycetes and strongly bound heavy metals (Clemençon, 1967, 1968, 1975b, 1978). Some of the small granules may be Golgi vesicles. The function of the siderophilous granules is largely unknown. This work needs to be repeated using aldehyde fixatives.

Evidence that lysosomes may exist in hymenia and gills of *Agaricus bisporus* has been obtained using biochemical methods and enzyme cytochemistry of tissues and cell fractions (Panzica and Panzica-Viglietti, 1974–1975; Scannerini et al., 1975).

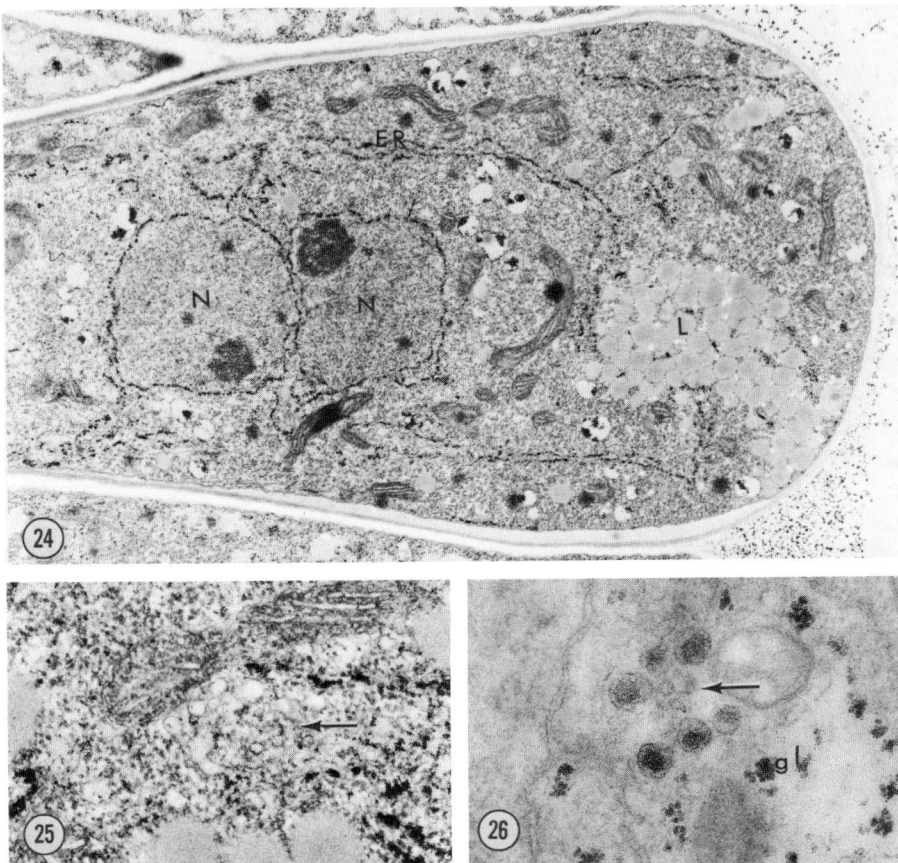

Figs. 2.24–2.26. Postmeiotic basidia of *Coprinus cinereus* after prolonged incubation in OsO$_4$ at 40 °C. Fig. 2.24 Basidium in longitudinal section with electron-dense deposits in endoplasmic reticulum (ER) and nuclear membrane. L, lipid droplet; N, nucleus (× 10,700). Fig. 2.25 Probable Golgi cisterna (arrow) without electron-dense deposits (× 39,500). Fig. 2.26 Golgi cisterna (arrow) in a postmeiotic basidium of *C. cinereus*, with vesicles stained with the silver-protein technique for carbohydrates. gl, glycogen (× 40,400).

Other Organelles

Cytoplasmic microtubules were longitudinally oriented at various stages in basidium and basidiospore development (McLaughlin, 1971, 1973, 1977a; Sundberg, 1977). A number of functions have been suggested, e.g., that they guide the movement of cell organelles into the forming spore. An experimental approach will be needed to understand their role in the basidium.

The relationship between microtubules, the spindle pole body, and migration of the nucleus into the basidiospore is unclear. In holobasidiate fungi the spindle pole body was apically positioned during nuclear migration through the sterigma (Setliff, 1977; Nakai and Ushiyama, 1978; Thielke, 1978). In some cases microtubules were reported to be connected to the spindle pole body (Nakai and Ushiyama, 1978; Thielke, 1978) but not in others (Setliff, 1977). In *Auricularia fuscosuccinea* (McLaughlin, unpubl. results) microtubules were not associated with the spindle pole body during nuclear migration into the sterigma. However, a mitotic division began in the sterigma and continued as the nucleus entered the spore. Astral microtubules were associated with the mitotic nucleus. The report of intranuclear microtubules during nuclear migration in *Lentinus edodes* (Nakai and Ushiyama, 1978) may indicate that the migrating nucleus was in mitosis, which has been suggested to be a common phenomenon (Wells, 1977). Microtubules are not always associated with the spindle pole body in migrating nuclei in basidiomycetes (Heath and Heath, 1978), and this association may not be essential for nuclear movement.

What changes mitochondria undergo as basidia form is not clear. The number of mitochondrial profiles increased during basidial maturation in *Agaricus campestris* Fr. (Manocha, 1965) and during basidiospore growth in *C. cinereus* (McLaughlin, 1977a). Microbodies have been identified by their morphology (McLaughlin, 1973) and need to be examined cytochemically. Our limited knowledge of the functions of microbodies, multivesicular bodies, and lomasomes in the basidiospore have recently been reviewed (McLaughlin, 1977a).

Experimental Control of Late Basidial Development

An intriguing problem in development is how the basidium can direct the positioning of typically 4, equally spaced outgrowths on its apex. The number of sterigmata on a basidium may vary from 1 to 8, and they are usually arranged in a regular pattern (Buller, 1922; Corner, 1972). Corner (1972) provided a geometrical analysis of the precisely determined basidiospore and sterigma positioning in holobasidia. The mechanism by which the basidium achieves this geometric precision is unknown. Pattern formation, in general, is a poorly understood process (Bonner, 1974).

Nuclear number and position have been suggested to influence sterigmal pattern. There appears to be a correlation between the number of sterigmata and the number of nuclei in basidia, since monokaryotic fruitbodies form 2-spored basidia, while their dikaryotic counterparts are 4-spored (Esser and Stahl, 1973; Esser et al., 1974). There is sometimes a correlation between the position of nuclei at the second division of meiosis and positioning of sterigmata (e.g., Gull and Newsam, 1976). However, the absence of a direct relationship between nuclear position at the second division of meiosis and sterigmal origin is suggested by the facts that sterigmata may arise before completion of meiosis (Sundberg, 1977) and that the position of nuclear division in a stichobasidium does not correlate with sterigmal origin.

Experiments on sterigmal initiation in *Coprinus cinereus* were begun to determine whether sterigmal pattern is fixed long before sterigma appear and whether it can be altered by the environment (McLaughlin, 1977b). Large numbers of sterigmata are initiated at about the same time in *C. cinereus*, making it a useful organism for such studies. Two experimental approaches were tried: chilling of intact fruitbodies and application of an electrical field to isolated gills. This research is still at an early stage, but it illustrates the potential of the gill of *C. cinereus* for experimental studies.

Chilling experiments were motivated by the chance observation that storage of cultures in the cold at −1 day seemed to cause abnormal sterigmal patterns. At the time of sterigmal initiation (Fig. 2.27), cultures were subjected to near freezing temperatures for 0–23 h and then returned to normal temperatures for 12 h.

Short chilling periods of up to 4 h had little effect on sterigmal patterns, but chilling for 24 h resulted in basidia bearing predominantly 1–3 sterigmata (Figs. 2.29–2.31). Fruitbody development was also affected by long cold treatments. Sterigmata were in regular patterns after chilling. If 1 sterigma was present, it was apically positioned; 2 sterigmata were opposite each other and 3 sterigmata were equidistant from each other. In control cultures not subjected to chilling treatment, the gill surface was obscured by spores in tetrads 12 h after sterigmal initiation (Fig. 2.28).

Results of chilling experiments suggest, first, that sterigmal pattern is determined at or just before sterigmata appear on basidia and, second, that sterigmal pattern is not fixed early in basidial development because it is subject to rearrangement when chilling is applied at the time of sterigmal initiation. This rearrangement is especially clear in apical positioning of single sterigmata.

An oriented electrical field was applied to isolated gills at the beginning of sterigmal initiation (Fig. 2.32) in the hope of shifting the positions of the sterigmata toward the positive or negative pole. Imposition of small steady electrical fields can induce oriented growth in *Fucus* eggs and other cells (Jaffe et al., 1974; Jaffe and Nuccitelli, 1977). The gill of *C. cinereus* is not ideal for these experiments because it is composed of several different cell types and because holobasidia cannot complete development if submerged (Corner, 1972).

In the electrical field all basidia that developed formed a single spore (Figs. 2.34, 2.35; Table 2.3). These basidiospores were sometimes giant and spherical (about 6 μm diameter; Fig. 2.34). Spores normally elongate before they reach this size. There was no clear orientation of single spores on treated gills to the positive or negative pole (Table 2.3). Maturation of basidia with 4 spores was suppressed in the electrical field. In the floating control (Fig. 2.33) basidia formed 4 spores. These spores were delayed in their development compared to those in the intact fruitbody (Table 2.3).

The electrical field inhibited all but one sterigma from developing. One possible explanation is that the electrical field overrides the orientation mechanism within the basidium which directs cell components to 4 growth points in the basidial apex and causes them to collect at a single site. Alternatively, the electrical field might exert its effect at the plasma membrane as has been demonstrated in *Micrasterias* (Brower and Giddings, 1980). Additional research will be needed to explain the results.

It is unclear how the basidium directs sterigmal initiation, but it may be via a system of chemical signals (Bonner, 1974). These signals must ultimately act in the cell cortex or at the plasma membrane. The primary effect of chilling has been

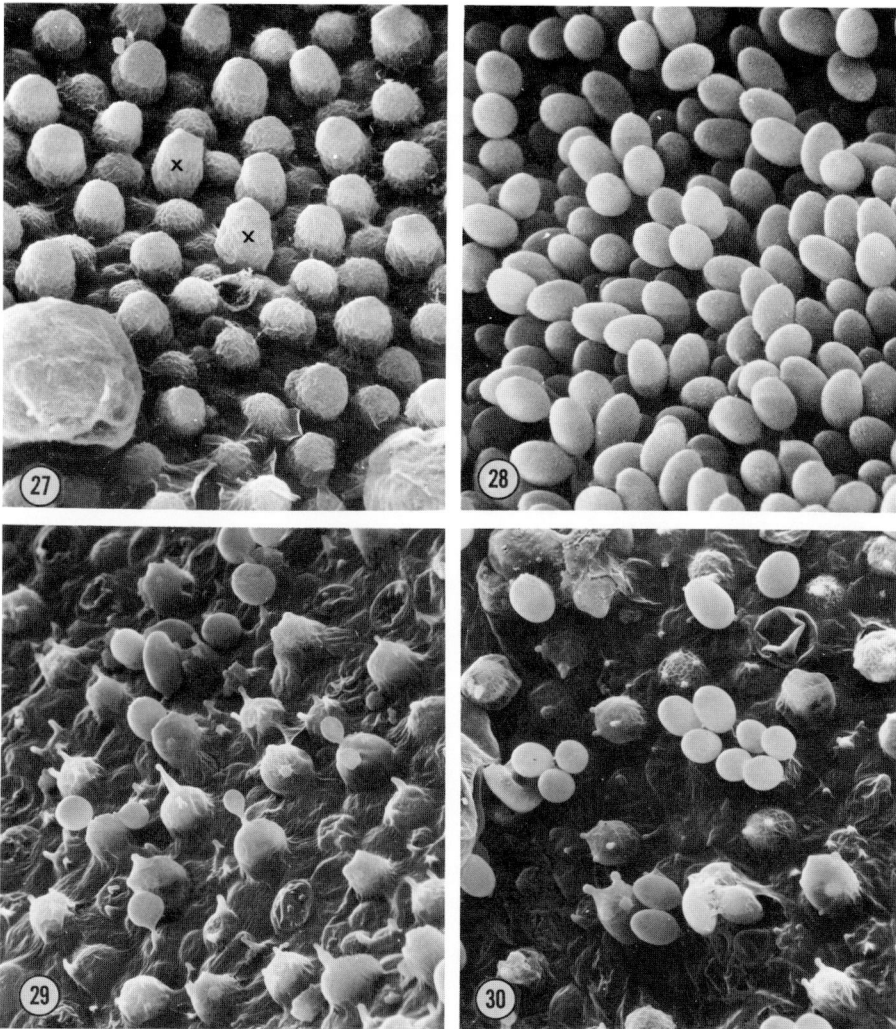

Figs. 2.27–2.30. Scanning electron micrographs of gills of *Coprinus cinereus* subjected to chilling at the time of sterigmata initiation or unchilled controls. Fig. 2.27 Gill surface at the start of the experiment with a few basidia (X) showing the first sign of sterigmata initiation, a squaring of the apex. Unchilled control ($\times$ 900). Fig. 2.28 Surface of a gill from the same fruitbody as in Fig. 2.27 but 12 h later. Basidiospores in tetrads mask the basidia. Unchilled control ($\times$ 850). Figs. 2.29, 2.30 Gills from two different experiments in which fruitbodies were chilled for 24 h and returned to normal temperatures for 12 h. Basidia with 1, 2, 3, or 4 sterigmata with or without spores. Basidia are dimorphic and sterigmata on different basidia are at several levels (Fig. 2.29, $\times$ 750; Fig. 2.30, $\times$ 750).

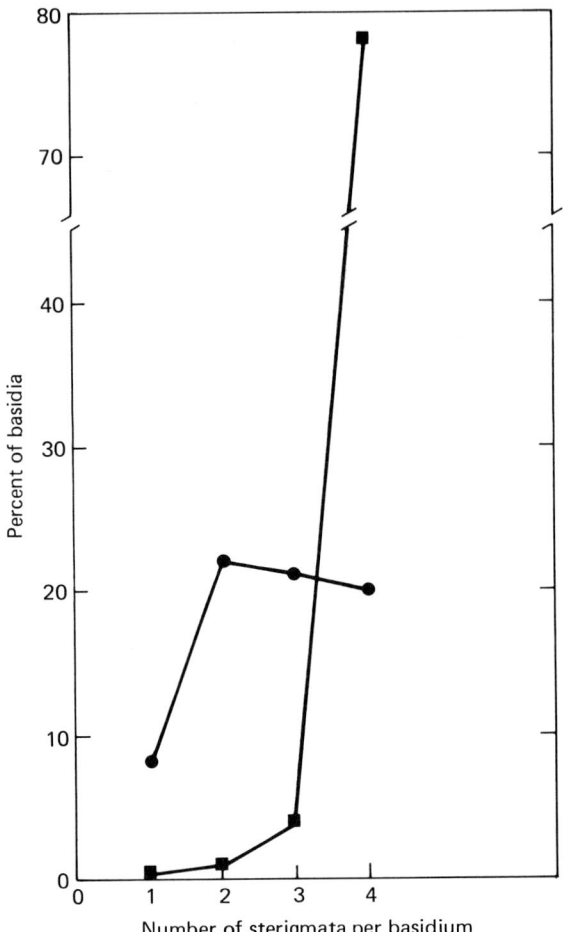

Fig. 2.31. Percentage of basidia of *Coprinus cinereus* bearing 1–4 sterigmata after 1–4 h chilling (■) or 24 h chilling (●).

suggested to be on cell membranes, but chilling also leads to numerous secondary effects (Lyons, 1973). Chilling is also known to depolymerize microtubules (Hepler and Palevitz, 1974), and it could affect a cytoskeletal or directive function that microtubules serve within the cell. An imposed electrical field may exert its effect at the plasma membrane (Jaffe and Nuccitelli, 1977).

Materials and Methods

Coprinus cinereus was grown in deep storage dishes as in McLaughlin (1974) on a 6 a.m.–6 p.m. day. Light microscopic cytochemical methods and controls are given in Table 2.1. The same procedures were used to localize protein and ribonucleic acid in the outer caps of the septal pore apparatus with the light microscope. For transmission electron microscopy the methods in McLaughlin (1977a)

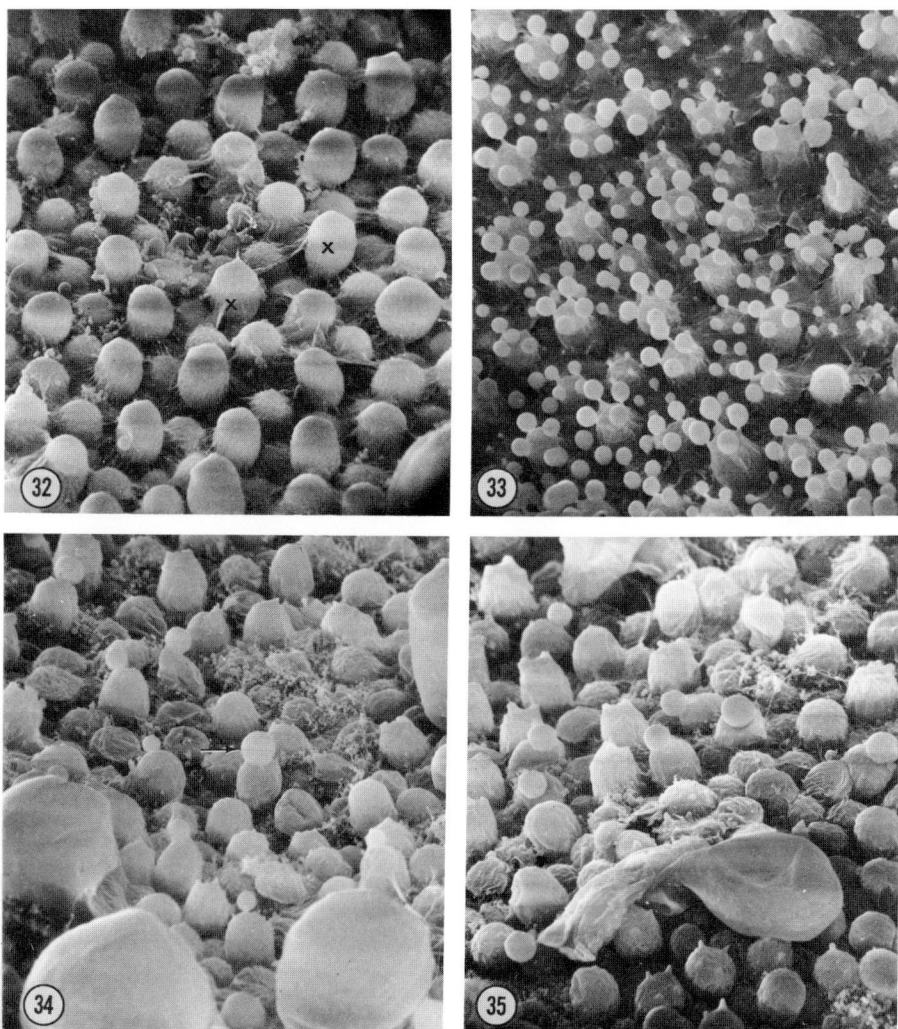

Figs. 2.32–2.35. Scanning electron micrographs of gills of *Coprinus cinereus* subjected to an electrical field at the time of sterigmata initiation and control gills. All gills from the same fruitbody. Fig. 2.32 Gill fixed at the start of the experiment. Basidia showing sterigmata initiation (X) ($\times$ 850). Fig. 2.33 Single gill floated in nutrient solution without the electrical field and fixed at the end of the experiment ($\times$ 750). Figs. 2.34, 2.35 Gill that developed in the electrical field. Basidia that have developed bear one spore which may be a giant and spherical (arrow) (Fig. 2.34, $\times$ 750; Fig. 2.35, $\times$ 850).

were followed except for Fig. 2.9, in which procedure 1 of McLaughlin (1974) was used. For high-voltage electron microscopy fixation 1 and embedding procedures in McLaughlin (1977a) were followed; sections, 0.25 μm thick, were collected on slotted grids using the technique of Rowley and Moran (1975), stained with 1.5% uranyl acetate in absolute methanol plus 1% dimethylsulfoxide for 20 min and then with lead citrate for 10 min. After staining, the grids were coated

Table 2.3. Electric Field Experiments with *Coprinus cinereus* Gill.[a]

	1-Spored basidia orientation			4-Sterigmate basidia		Basidia without sterigmata or spores	Number of basidia counted
	+	−	±	Stage[b] 0–2 spores	Stage 3–4 spores		
Experiment 1							
Start				26.2		73.8	130
End							
treatment[c]	6.5	13.4	3.4	47.6		28.5	321
floating control				77.4	11.9	9.0	177
intact fruitbody				12.8	87.2	0	39
Experiment 2							
Start				9.2		90.8	109
End							
treatment[c]	7.8	1.1	1.6	10.5		75.8	372
floating control				81.9	11.0	7.1	155
intact fruitbody					97.1	2.9	35

[a]Results expressed as % of basidia on gill giving this response.
[b]For 4 stages in basidiospore development see McLaughlin (1977a).
[c]Treatment: gill that developed in the electrical field. Floating control: gill that developed in a parallel situation to treated gill but without the electrical field. Intact fruitbody: gill that developed on the fruitbody.

on both sides with a layer of carbon and the sections examined in a JEM 1000 electron microscope at 1000 kV. Carbohydrates were localized in the electron microscope with the silver-protein technique (McLaughlin, 1974). For osmium impregnation of the endomembrane system gills were fixed in 4% glutaraldehyde with 0.02 M sucrose and 0.0001 M $CaCl_2$ in 0.1 M sodium cacodylate buffer, pH 7.2, 1 h, room temperature, then incubated at 40 °C in unbuffered 2% OsO_4 for 40 h with a solution change after 24 h. Tissue was embedded and sectioned as in McLaughlin (1977a). Sections were stained 5 min in aqueous uranyl acetate and 15 min in lead citrate (Fig. 2.24) or unstained (Fig. 2.25). For freeze-etch preparations gills were fixed in 4% glutaraldehyde as above, infiltrated with glycerol (20%) for 1.5–3 h, fractured at −100 °C in a Balzers freeze-etching instrument, etched for 2 min or longer, and the replicas cleaned with bleach and nitric acid.

The procedure for chilling experiments on *C. cinereus* was as follows: fruitbodies were produced on yeast–malt–glucose agar (Heintz and Niederpruem, 1971) in deep storage dishes (McLaughlin, 1974). When sterigmal initiation could be detected on a few basidia with the light microscope, cultures were placed in an ice bath for 1 h, then at 5 °C for 0–23 h. In some cases only 5 °C was used. After chilling, cultures were returned to normal growth conditions for 12 h. A few gills can be removed from the fruitbody without interfering with its development. Some gills were fixed in buffered glutaraldehyde (McLaughlin, 1977a) at the start of the experiment, at the end of chilling treatment, and 12 h later. Distribution of basidia with different numbers of sterigmata was determined at selected sites on the gill face with light or scanning electron microscopy.

The procedure for electrical field experiments with *C. cinereus* was as follows: at the time of sterigmal initiation (Fig. 2.32) single gills were peeled from slices of the pileus and maintained at the surface of a nutrient solution on two plastic supports in a petri dish. Madelin's (1956) medium, ⅕ strength, buffered with 6 mM BES was used. The apparatus consisted of two outer wells filled with 4 mM KCl through which current was passed via salt bridges to the central petri dish containing the gill. A multimeter connected to platinum electrodes measured the current within the center of the petri dish. An electric current, 12.5–13 V/cm, was passed across the long axis of the gill. Fresh solution (about 900 ml/h) flowed through the petri dish parallel to the long axis of the gill. As in the chilling experiments some gills were fixed at the start of the experiment. The experiments were terminated when inspection of gills on the fruitbodies showed advanced stages in spore formation. Experiments usually lasted 5 h or longer. The floating control was a single gill maintained in the nutrient solution without an electrical current; the solution was not flowing in this control and the pH remained constant. Gills were fixed at the end of the experiment and critical-point dried. Selected sites on the gill were examined with a scanning electron microscope.

Fruitbodies of *Boletus rubinellus* were obtained on Hagem's agar as in McLaughlin (1971). For electron microscopy pieces of the hymenium were fixed in 4% glutaraldehyde in 0.1 M sodium cacodylate buffer, pH 7.2, for 1 h at room temperature, postfixed in 1% OsO_4 in the same buffer for 4 h at room temperature, stained in 0.5% aqueous uranyl acetate for 2.5 h, dehyrated in acetone, and epon-embedded.

Fruitbodies of *Auricularia fuscosuccinea* were obtained as in McLaughlin (1979). For electron microscopy pieces of the hymenium were fixed in 3% glutaraldehyde in 0.1 M phosphate buffer, pH 7.2, for 15 min at room temperature, then an equal volume of 2% OsO_4 in the same buffer was added and fixation continued at 4 °C for an additional hour. Tissue was stained overnight in 0.5% aqueous uranyl acetate, dehydrated, embedded, serial-sectioned, and examined as in Yoon and McLaughlin (1979).

Summary and Conclusions

Changes in cytoplasm and wall in holobasidia and phragmobasidia are reviewed with major emphasis on basidia of *Coprinus cinereus*. Before karyogamy the basidium of *C. cinereus* contained few types of organelles and the apex did not resemble the growing tip of a vegetative hypha; however, vesicles have been found at the apices of promycelia. Light microscopic cytochemistry showed that intense staining for ribonucleic acids and proteins began before karyogamy in *C. cinereus*. The type and timing of storage product formation varied with the species. In *C. cinereus* the basidium wall resembled that of subhymenial cells and not those of other hymenial elements. A trilaminar pellicle and a rodlet pattern enveloped the entire surface of the gill and appeared to enclose the basidiospore. The sterigmal wall consisted of a single layer and was either a modification of the basidial wall or a new wall layer inserted into the basidial apex. Teliospore and basidiospore development are reviewed. Septation within phragmobasidia was compared and a new type demonstrated for *Auricularia fuscosuccinea*.

An outer cap of differentiated cytoplasm surrounded the septal pore cap at the basidial base in many holobasidiate fungi. The "ampoule effect" (Corner, 1948) was re-evaluated and found to be an unsatisfactory model of the cellular processes involved in basidiosporogenesis. The morphology of the Golgi cisternae in *C. cinereus* was determined using serial sections for standard and high-voltage electron microscopy, and 3 basic types were found: "ring" or sphere, plate, and tubular-vesicular. Attempts to identify the initial form of the Golgi cisternae with long incubation in OsO_4 resulted only in staining within the endoplasmic reticulum and nuclear membrane. The contribution of other cell components to basidial development are reviewed.

Two experimental approaches to control sterigma and basidiospore development in *C. cinereus* were tried: application of a chilling treatment to intact fruitbodies and of an electrical field to individual gills developing *in vitro*. Both treatments were applied at the start of sterigmal initiation. After long chilling followed by 12 h growth at normal temperatures, basidia bore predominantly 1, 2, and 3 sterigmata arranged in regular patterns. Application of an oriented electrical field to floating gills resulted in basidia with one immature spore. The electrical field did not induce a shift in the site of sterigmal initiation.

While cytological information on meiotic and postmeiotic events in holobasidia is now accumulating, little information is available on premeiotic differentiation. There is a need for detailed studies of phragmobasidial development comparable to those on holobasidia to reveal which aspects of differentiation of the cytoplasm and walls are fundamental to all basidia. Such studies may contribute to an understanding of the basidiospore as a ballistospore, the hallmark of the basidiomycetes. Further ultrastructural, cytochemical, and other experimental approaches to basidial development will help determine the function of various organelles and the significance of such events as "charging," and of differentiation of the walls of sterigmata and basidiospores for the development process. Gills of *Coprinus cinereus* appear to be suitable for use in an *in vitro* system for the study of late stages in basidial development.

Acknowledgments. I thank Allen Seilheimer and Kwon Yoon for the use of micrographs, Dan Brower for assistance and advice in the electrical field experiments, and Robert Nowak for assistance with the chilling experiments. I thank Jeremy Pickett-Heaps for providing laboratory space, Iris Charvat and Kerry O'Donnell for helpful discussions, and Esther McLaughlin for editing.

The research was supported by N.S.F. Grant BMS 72-02516 and DEB 78-23392 and a grant from the Graduate School, University of Minnesota. High-voltage electron microscopy was supported by Grant RR-00592 from the Division of Research Resources, National Institutes of Health.

References

Allen, J. V., Hess, W. M., Weber, D. J.: Ultrastructural investigations of dormant *Tilletia caries* teliospores. Mycologia *63*, 144–156 (1971).
Ashton, M. L., Moens, P. B.: Ultrastructure of sporulation in the Hemiascomycetes *Ascoidea corymbosa, A. rubescens, Cephaloascus fragrans*, and *Saccharomyces capsularis*. Can. J. Bot. *57*, 1259–1284 (1979).
Bartnicki-Garcia, S.: Fundamental aspects of hyphal morphogenesis. Symp. Soc. Gen. Microbiol. *23*, 245–267 (1973).
Beckett, A., Heath, I. B., McLaughlin, D. J.: An Atlas of Fungal Ultrastructure. London: Longmans Group Ltd. 1974.
Beever, R. E., Dempsey, G. P.: Function of rodlets on the surface of fungal spores. Nature, Lond. *272*, 608–610 (1978).
Bennell, A. R., Henderson, D. M.: Urediniospore and teliospore development in *Tranzschelia* (Uredinales). Trans. Br. Mycol. Soc. *71*, 271–278 (1978).
Bennell, A. R., Henderson, D. M., Prentice, H. T.: The teliospores of *Puccinia smyrnii* (Uredinales). Grana *17*, 17–27 (1978).
Bonner, J. T.: On Development. Cambridge: Harvard University Press. 1974.
Bracker, C. E.: Ultrastructure of fungi. Annu. Rev. Phytopathol. *5*, 343–374 (1967).
Bracker, C. E., Ruiz-Herrera, J., Bartnicki-Garcia, S.: Structure and transformation of chitin synthetase particles (chitosomes) during microfibril synthesis *in vitro*. Proc. Natl. Acad. Sci. U.S.A. *73*, 4570–4574 (1976).
Bronchart, R., Demoulin, V.: Ultrastructure de la paroi des basidiospores de *Lycoperdon* et de *Scleroderma* (Gastéromycètes) comparée à celle de quelques autres spores de champignon. Protoplasma *72*, 179–189 (1971).
Bronchart, R., Calonge, F. D., Demoulin, V.: Nouvelle contribution a l'etude de l'ultrastructure de la paroi sporale des Gastéromycètes. Bull. Soc. Mycol. Fr. *91*, 231–246 (1975).
Brower, D. L., Giddings, T. H.: The effects of applied electric fields on *Micrasterias*. II. The distribution of cytoplasmic and plasma membrane components. J. Cell Sci. *42*, 279–290 (1980).
Buller, A. H. R.: Researches on Fungi, Vol. 2. London: Longmans, Green and Co. 1922.
Burge, H. A.: Basidiospore structure and development in the genus *Russula*. Mycologia *71*, 977–995 (1979).
Campbell, R.: Ultrastructure of asci, ascospores, and spore release in *Lophodermella sulcigena* (Rostr.) v. Hohn. Protoplasma *78*, 69–80 (1973).
Clemençon, H.: Beiträge zur Kenntnis der Gattungen *Lyophyllum* und *Calocybe*. II. Cytochemie und Feinstruktur der Basidie von *Lyophyllum urbanese* spec. nov. Nova Hedwigia *14*, 127–142 (1967).
Clemençon, H.: Beiträge zur Kenntnis der Gattungen *Lyophyllum* und *Calocybe*. (Agaricales, Basidiomycetes). VI. Die Entwicklung der siderophilen Granulation. Cytologia (Tokyo) *33*, 498–507 (1968).
Clemençon, H.: Reifung und endoplasmatisches Retikulum der Agaricales-Basidie. Z. Pilzkd. *35*, 295–304 (1969).
Clemençon, H.: Die Wandstrukturen der Basidiosporen. V. *Pholiota* und *Kuehneromyces*, verglichen mit *Galerina* und *Gymnopilus*. Z. Pilzkd. *40*, 105–126 (1974).
Clemençon, H.: Electron microscopy and spore teguments. In: The Agaricales in Modern Taxonomy, 3rd ed., Singer, R. (ed.). Vaduz: J. Cramer 1975a, pp. 83–89.

Clemençon, H.: Ultrastructure of hymenial cells in two boletes. Nova Hedwigia 51, 93–98 (1975b).
Clemençon, H.: Die Strukturen der Basidiosporenwand und des Apikulus, und deren Beziehung zur Exogenisation der Spore. Persoonia 9, 363–380 (1977).
Clemençon, H.: Siderophilous granules in the basidia of Hymenomycetes. Persoonia 10, 83–96 (1978).
Coffey, M. D.: Obligate parasites of higher plants particularly rust fungi. Symp. Soc. Exp. Biol. 29, 297–323 (1975).
Corner, E. J. H.: Studies in the basidium. I. The ampoule effect, with a note on nomenclature. New Phytol. 47, 22–51 (1948).
Corner, E. J. H.: Studies in the basidium. Spore-spacing and the Boletus spore. Gard. Bull. (Singapore) 26, 159–194 (1972).
Craig, G. D., Newsam, R. J., Gull, K.: Subhymenial branching and dolipore septation in Agaricus bisporus. Trans. Br. Mycol. Soc. 69, 337–344 (1977).
Donk, M. A.: The Heterobasidiomycetes: A reconnaissance-II. Some problems connected with the restricted emendation. Proc. K. Ned. Akad. Wet. Ser. C Biol. Med. Sci. 75, 376–390 (1972).
Esser, K., Semerdžieva, M., Stahl, U.: Genetische Untersuchungen an den Basidiomyceten Agrocybe aegerita I. Eine Korrelation zwischen dem Zeitpunkt der Fruchtkörperbildung und monokaryotischem Fruchten und ihre Bedeutung für Züchtung und Morphogenese. Theor. Appl. Genet. 45, 77–85 (1974).
Esser, K., Stahl, U.: Monokaryotic fruiting in the basidiomycete Polyporus ciliatus and its suppression by incompatibility factors. Nature, Lond. 244, 304–305 (1973).
Feder, N., O'Brien, T. P.: Plant microtechnique: Some principles and new methods. Am. J. Bot. 55, 123–142 (1968).
Fisher, D. B.: Protein staining of ribboned epon sections for light microscopy. Histochemie 16, 92–96 (1968).
Flegler, S. L., Hooper, G. R., Fields, W. G.: Ultrastructural and cytochemical changes in the basidiomycete dolipore septum associated with fruiting. Can. J. Bot. 54, 2243–2253 (1976).
Friend, D. S.: Cytochemical staining of multivesicular body and Golgi vesicles. J. Cell Biol. 41, 269–279 (1969).
Friend, D. S., Murray, M. J.: Osmium impregnation of the Golgi apparatus. Am. J. Anat. 117, 135–150 (1965).
Garcia Mendoza, C., Leal, J. A., Novaes-Ledieu, M.: Studies of the spore walls of Agaricus bisporus and Agaricus campestris. Can. J. Microbiol. 25, 32–39 (1979).
Gardner, J. S., Allen, J. V., Hess, W. M.: Fixation of dormant Tilletia teliospores for thin sectioning. Stain Technol. 50, 347–350 (1975).
Gardner, J. S., Hess, W. M.: Ultrastructure of lipid bodies in Tilletia caries teliospores. J. Bacteriol. 131, 662–671 (1977).
Girbardt, M.: Ultrastructure and dymamics of the moving nucleus. Symp. Soc. Exp. Biol. 22, 249–259 (1968).
Girbardt, M., Hädrich, H.: Ultrastruktur des Pilzkernes III. Genese des kernassoziierten Organells (NAO = "KCE"). Z. Allg. Mikrobiol. 15, 157–173 (1975).
Greuter, B., Rast, D.: Ultrastructure of the dormant Agaricus bisporus spore. Can. J. Bot. 53, 2096–2101 (1975).
Grove, S. N., Bracker, C. E.: Protoplasmic organization of hyphal tips among fungi: Vesicles and Spitzenkörper. J. Bacteriol. 104, 989–1009 (1970).
Gull, K.: Cytoplasmic microfilament organization in two basidiomycete fungi. J. Ultrastruct. Res. 50, 226–232 (1975).

Gull, K.: Differentiation of septal ultrastructure according to cell type in the basidiomycete, *Agrocybe praecox*. J. Ultastruct. Res. *54*, 89–94 (1976).
Gull, K., Newsam, R. J.: Meiosis in the basidiomyceteous fungus, *Coprinus atramentarius*. Protoplasma *90*, 343–352 (1976).
Harder, D. E.: Electron microscopy of teliospore formation in *Puccinia coronata avenae*. Physiol. Plant Pathol. *10*, 21–28 (1977).
Hashimoto, T., Wu-Yuan, C. D., Blumenthal, H. J.: Isolation and characterization of the rodlet layer of *Trichophyton mentagrophytes* microconidial wall. J. Bacteriol. *127*, 1543–1549 (1976).
Heath, I. B., Heath, M. C.: Microtubules and organelle movements in the rust fungus *Uromyces phaseoli* var. *vignae*. Cytobiologie *16*, 393–411 (1978).
Heintz, C. E., Niederpruem, D. J.: Ultrastructure of quiescent and germinated basidiospores and oidia of *Coprinus lagopus*. Mycologia *63*, 745–766 (1971).
Henderson, D. M., Eudall, R., Prentice, H. T.: Morphology of the reticulate teliospores of *Puccinia chaerophylli*. Trans. Br. Mycol. Soc. *59*, 229–232 (1972).
Hepler, P. K., Palevitz, B. A.: Microtubules and microfilaments. Annu. Rev. Plant Physiol. *25*, 309–362 (1974).
Hess, W. M., Bushnell, J. L., Weber, D. J.: Surface structures and unidentified organelles of *Lycoperdon perlatum* Pers. basidiospores. Can. J. Microbiol. *18*, 270–271 (1972).
Hess, W. M., Weber, D. J.: Form and function in basidiomycete spores. In: The Fungal Spore. Weber, D. J., Hess, W. M. (eds.) New York: John Wiley & Sons 1976, pp. 643–713.
Hoch, H. C., Setliff, E. C.: Sterigma and basidiospore development in *Poria latemarginata*. Mem. N. Y. Bot. Gard. *28*, 98–104 (1976).
Howard, R. J., Aist, J. R.: Hyphal tip cell ultrastructure of the fungus *Fusarium*: improved preservation by freeze-substitution. J. Ultrastruct. Res. *66*, 224–234 (1979).
Hugueney, R.: Ontogenèse des infrastructures de la paroi sporique de *Coprinus cineratus* Quél. var. *nudisporus* Kühner (Agaricales). C. R. Hebd. Seances Acad. Sci., Ser. D. *275*, 1495–1498 (1972).
Hugueney, R.: Morphologie, ultrastructure et developpement de l'apicule des spores de quelques Coprinacées: étude particulière du punctum lacrymans. Bull. Soc. Linn. Lyon *44*, 249–256 (1975).
Jaffe, L. S., Nuccitelli, R.: Electrical controls of development. Annu. Rev. Biophys. Bioeng. *6*, 445–476 (1977).
Jaffe, L. S., Robinson, K. R., Nuccitelli, R.: Local cation entry and self-electrophoresis as an intracellular localization mechanism. Ann. N.Y. Acad. Sci. *238*, 372–389 (1974).
Jensen, W. A.: Botanical Histochemistry. San Francisco: W. H. Freeman 1962.
Keller, J.: Contribution à la connaissance de l'infrastructure de la paroi sporique des Aphyllophorales. Doctoral Thesis, Univ. Neuchatel. 1974.
Keller, J.: Ultrastructure des parois sporiques des Aphyllophorales. IV. Ontogenese des parois sporiques de *Pachykytospora tuberculosa* et de *Ganoderma lucidum*. Bull. Soc. Bot. Suisse *87*, 34–51 (1977).
Khan, S. R., Kimbrough, J. W.: Ultrastructure of septal pore apparatus in the lamellae of *Nematoloma puiggarii*. Can. J. Bot. *57*, 2064–2070 (1979).
Khan, S. R., Talbot, P. H. B.: Ultrastructure of septa in hyphae and basidia of *Tulasnella*. Mycologia *68*, 1027–1036 (1976).
Kohno, M., Nishimura, T., Ishizaki, H., Kunoh, H.: Ultrastructural changes of cell wall in germinating teliospore of *Gymnosporangium haraeanum* Sydow. Trans. Mycol. Soc. Jpn. *16*, 106–112 (1975).

Kollmorgen, J. F., Hess, W. M., Trione, E. J.: Ultrastructure of primary sporidia of wheat-bunt fungus, *Tilletia caries*, during ontogeny and mating. Protoplasma *99*, 189–202 (1979).
Kühner, R.: Architecture de la paroi sporique des hyménomycètes et de ses differénciations. Persoonia *7*, 217–248 (1973).
Lerbs, V.: Licht- und elektronenmikroskopische Untersuchungen an meiotischen Basidien von *Coprinus radiatus* (Bolt.) Fr. Arch. Mikrobiol. *77*, 308–330 (1971).
Littlefield, L. J., Heath, M. C.: Ultrastructure of Rust Fungi. New York: Academic Press 1979.
Lyons, J. M.: Chilling injury in plants. Annu. Rev. Plant Physiol. *24*, 445–466 (1973).
Madelin, M. F.: The influence of light and temperature on fruiting of *Coprinus lagopus* Fr. in pure culture. Ann. Bot. (Lond.) *20*, 467–480 (1956).
Manocha, M. S.: Fine structure of the *Agaricus* carpophore. Can. J. Bot. *43*, 1329–1333 (1965).
McLaughlin, D. J.: Some aspects of hymenial fine structure in the mushroom *Boletus rubinellus*. Am. J. Bot. *57*, 745 (1970).
McLaughlin, D. J.: Centrosomes and microtubules during meiosis in the mushroom *Boletus rubinellus*. J. Cell Biol. *50*, 737–745 (1971).
McLaughlin, D. J.: Golgi apparatus in the postmeiotic basidium of *Coprinus lagopus*. J. Bacteriol. *109*, 739–742 (1972).
McLaughlin, D. J.: Ultrastructure of sterigma growth and basidiospore formation in *Coprinus* and *Boletus*. Can. J. Bot. *51*, 145–150 (1973).
McLaughlin, D. J.: Ultrastructural localization of carbohydrates in the hymenium of *Coprinus*. Evidence for the function of the Golgi apparatus. Protoplasma *82*, 341–364 (1974).
McLaughlin, D. J.: Basidiospore initiation and early development in *Coprinus cinereus*. Am. J. Bot. *64*, 1–16 (1977a).
McLaughlin, D. J.: Ultrastructure and cytochemistry of basidial and basidiospore development. Second Int. Mycol. Congress, Tampa, Fl., Abstracts, p. 433 (1977b).
McLaughlin, D. J.: Ultrastructure of the hymenium of *Auricularia polytricha*. Mushroom Sci. *10*, 219–229 (1979).
McLaughlin, D. J. Ultrastructure of the metabasidium of *Auricularia fuscosuccinia*. Am. J. Bot. *67*, 1225–1235 (1980).
McLaughlin, D. J., McLaughlin, E. G.: Anatomy and fine structure of the dimitic basidiomycete *Pterula*. Second Int. Mycol. Congress, Tampa, Fl., Abstracts, p. 434 (1977).
Mims, C. W.: Ultrastructure of teliospore formation in the cedar-apple rust fungus *Gymnosporangium juniperi-virginianae*. Can. J. Bot. *55*, 2319–2329 (1977).
Mims, C. W., Seabury, F., Thurston, E. L.: Fine structure of teliospores of the cedarapple rust *Gymnosporangium juniperi-virginianae*. Can. J. Bot. *53*, 544–552 (1975).
Mims, C. W., Thurston, E. L.: Ultrastructure of teliospore formation in the rust fungus *Puccinia podophylli*. Can. J. Bot. *57*, 2533–2538 (1979).
Morré, D. J., Mollenhauer, H. H.: The endomembrane concept: A functional integration of endoplasmic reticulum and Golgi apparatus. In: Dynamic Aspects of Plant Ultrastructure. Robards, A. W. (ed.). London: McGraw-Hill 1974, pp. 84–137.
Müller, L. Y., Rijkenberg, F. H. J., Truter, S. J.: A preliminary ultrastructural study on the *Uromyces appendiculatus* teliospore stage. Phytophylactica *6*, 123–128 (1974).
Nakai, Y., Ushiyama, R.: Fine structure of shiitake, *Lentinus edodes* (Berk.) Sing. II. Development of basidia and basidiospores. Rep. Tottori Mycol. Inst. *11*, 7–15 (1974).
Nakai, Y., Ushiyama, R.: Fine structure of shiitake, *Lentinus edodes*. VI. Cytoplasmic microtubules in relation to nuclear movement. Can. J. Bot. *56*, 1206–1211 (1978).

Oláh, G. M.: The fine structure of *Psilocybe quebecensis*. Mycopathol. Mycol. Appl. *49*, 321–338 (1973).
Oláh, G. M., Cole, G. T., Reisinger, O.: La role et la nature chimique des microvésicules sécrétories dans l'apex hyphal et dans les cellules sporogenes. Ann. Sci. Nat. Bot. Biol. Vég. Ser. 12, *18*, 301–318 (1977).
Oláh, G. M., Reisinger, O.: L'ontogenie des téguments de la paroi sporale en relation avec la stérigmate et la gouttelette hilaire chez quelques agarics melanosporés. C. R. Hebd. Seances Acad. Sci., Ser. D *278*, 2755–2758 (1974).
Panzica, G., Panzica-Viglietti, C.: Cytochemical technique for subcellular acid phosphatase in cultivated mushroom "*Psalliota bispora*" Quel. fractions. Allionia (Turin) *20*, 19–22 (1974–75).
Pegler, D. N., Young, T. W. K.: Ultrastructure of basidiospores in Agaricales in relation to taxonomy and spore discharge. Trans. Br. Mycol. Soc. *52*, 491–496 (1969).
Pegler, D. N., Young, T. W. K.: Basidiospore morphology in the Agaricales. Nova Hedwigia *35*, 1–210 (1971).
Pegler, D. N., Young, T. W. K.: The gasteroid Russulales. Trans. Br. Mycol. Soc. *72*, 353–388 (1979).
Perreau-Bertrand, J.: Recherches sur la différenciation et la structure de la paroi sporale chez les Homobasidiomycètes à spores ornées. Ann. Sci. Nat. Bot. Biol. Vég. 12, *8*, 639–746 (1967).
Perreau, J.: Développement, morphologie et structure de la basidiospore (chez les Homobasidiomycètes). L'Information Scientifique *31*, 55–75 (1976).
Perreau, J.: A propos de l'appendice hilaire des basidiospores: Organisation de la partie proximale sporique chez quelques Gastéromycètes. Rev. Mycol. (Paris) *41*, 363–379 (1977).
Ramberg, J. E.: Ultrastructural Study of Promycelial Development and Basidiospore Initiation in *Ustilago maydis*. M. Sc. thesis, University of Minnesota. 1979.
Rast, D., Hollenstein, G. O.: Architecture of the *Agaricus bisporus* spore wall. Can. J. Bot. *55*, 2251–2262 (1977).
Robb, J.: Ultrastructure of *Ustilago hordei*. I. Pregermination development of hydrating teliospores. Can. J. Bot. *50*, 1253–1261 (1972a).
Robb, J.: Ultrastructure of *Ustilago hordei* (Pers.) Lagerh. II. Septation in the metabasidium. Can. J. Genet. Cytol. *14*, 839–849 (1972b).
Rowley, J. C., Moran, D. T.: A single procedure for mounting wrinkle free sections on formvar coated slot grids. Ultramicrotomy *1*, 151–155 (1975).
Savile, D. B. O.: Possible interrelationships between fungal groups. In: The Fungi, Vol. 3. Ainsworth, G. C., Sussman A. S. (eds.). New York: Academic Press 1968, pp. 649–675.
Scannerini, S., Guinta, C., Panzica-Viglietti, C., Panzica, G.: Lysosomes in the cultivated mushroom (*Psalliota bispora* Quel.). G. Batteriol. Virol. Immunol. *68*, 1–15 (1975).
Setliff, E. C.: Ultrastructural studies of *Phanerochaete chrysosporium*. II. Nuclear migration through a sterigma. Second Int. Mycol. Congress, Tampa, Fl., Abstracts, p. 608 (1977).
Setliff, E. C., Hoch, H. C., Patton, R. F.: Studies on nuclear division in basidia of *Poria latemarginata*. Can. J. Bot. *52*, 2323–2333 (1974).
Sundberg, W. J.: Hymenial cytodifferentiation in basidiomycetes. In: The Filamentous Fungi, Vol. 3. Smith, J. E., Berry, D. R. (eds.). London: Edward Arnold 1977, pp. 298–314.
Syrop, M. J., Beckett, A.: The origin of ascospore-delimiting membranes in *Taphrina deformans*. Arch. Mikrobiol. *86*, 185–191 (1972).

Thielke, C.: Die Feinstruktur der Basidien des Kulturchampignons. Arch. Mikrobiol. *59*, 405–407 (1967).
Thielke, C.: Membransysteme in meiotischen Basidien. Ber. Dtsch. Bot. Ges. *81*, 183–186 (1968a).
Thielke, C.: Restitution der Kernmembran in postmeiotischen Basidien. Ber. Dtsch. Bot. Ges. *81*, 315–316 (1968b).
Thielke, C.: Die Substruktur der Zellen im Fruchtkörper von *Psalliota bispora*. Mushroom Sci. *7*, 23–30 (1969).
Thielke, C.: Die Dolipore der Basidiomyceten. Arch. Mikrobiol. *82*, 31–37 (1972).
Thielke, C.: Intranucleäre Spindeln und Reduktion des Kernvolumens bei der Meiose von *Coprinus radiatus* (Bolt) Fr. Arch. Microbiol. *98*, 225–237 (1974).
Thielke, C.: Feinstrukturen bei Basidiomyceten. Z. Mykol. *44*, 71–89 (1978).
Tsuneda, I., Kennedy, L. L.: Ultrastructure of basidiospore germination in *Fomes fomentarius*. Can. J. Bot. *56*, 2865–2872 (1978).
Tu, C. C., Kimbrough, J. W., Aldrich, H. C.: Cytology and ultrastructure of *Thanatephorus cucumeris* and related taxa of the *Rhizoctonia* complex. Can. J. Bot. *55*, 2419–2436 (1977).
van der Valk, P., Wessels, J. G. H.: Ultrastructure and localization of wall polymers during regeneration and reversion of protoplasts of *Schizophyllum commune*. Protoplasma *90*, 65–87 (1976).
Wells, K.: The basidia of *Exidia nucleata*. I. Ultrastructure. Mycologia *56*, 327–341 (1964a).
Wells, K.: The basidia of *Exidia nucleata*. II. Development. Am. J. Bot. *51*, 360–370 (1964b).
Wells, K.: Ultrastructural features of developing and mature basidia and basidiospores of *Schizophyllum commune*. Mycologia *57*, 236–261 (1965).
Wells, K.: Meiotic and mitotic divisions in the Basidiomycotina. In: Mechanisms and Control of Cell Division. Rost, T. L., Gifford, E. M., Jr. (eds.). Stroudsburg, Pa.: Dowden, Hutchinson and Ross 1977, pp. 337–374.
Wells, K.: The fine structure of septal pore apparatus in the lamellae of *Pholiota terrestris*. Can. J. Bot. *56*, 2915–2924 (1978).
Yoon, K. S., McLaughlin, D. J.: Formation of the hilar appendix in basidiospores of *Boletus rubinellus*. Am. J. Bot. *66*, 870–873 (1979).

Chapter 3

Meiotic Divisions in the Basidium

CHARLOTTE THIELKE

Introduction

The main function of meiotic divisions is the reduction and recombination of chromosomal material. Evidence from genetic research seems to indicate that in the higher fungi these processes occur in a rather conventional manner (Esser and Kuenen, 1967). Concerning the morphogenesis of meiosis, however, there are a number of unresolved questions. These questions have been partially answered by several recent papers (e.g., Lu, 1966a, 1966b, 1967; McLaughlin, 1971; Setliff et al., 1974; Gull and Newsam, 1976; Wells, 1977). A review of the literature illustrates that there are different opinions with respect to the morphology of the spindle apparatus and associated structures and their behavior during meiosis. Therefore, it would seem to be useful to discuss these structures and their possible functions about which there is disagreement. The major topics are the nuclear envelope, the morphology of the chromatin, and the kinetic apparatus.

Nuclear division in higher fungi has some characteristics that distinguish it from the better-known nuclear division in higher plants. Meiotic basidia of Homobasidiomycetes are favorable objects for the study of nuclear division because of their regular orientation of the division spindles. This report refers to about 20 different species of Agaricales: i.e., *Agaricus bisporus* (Lange) Imbach, *Coprinus cinereus* (Schaeff. ex Fr.) S. F. Gray, *C. micaceus* (Bull. ex Fr.) Fr., *C. radiatus* (Bolt. ex Fr.) S. F. Gray, *Flammulina velutipes* (Curt. ex Fr.) Sing., *Laccaria amethystina* (Bolt. ex Hooker) Murr., *Lentinus edodes* (Berk.) Sing., *Lepista nuda* (Bull. ex Fr.) Cooke, *Panaeolina foenisecii* (Pers. ex Fr.) Maire, *Pleurotus ostreatus* (Jacq. ex Fr.) Kummer, *Psilocybe turficola* Favre, *Entoloma saundersii* (Fr.) Sacc., *Stropharia rugosoannulata* Farlow apud Murr. We preferred the species of *Coprinus* because in this species the hymenia develop in a relatively synchronous manner. From our experience it appears that within the Agaricales the mechanics of nuclear divisions are essentially similar. This seems also to be the case in *Poria latemarginata* (Dur. et Mont.) Cooke (Setliff et al., 1974) and in *Boletus rubinellus* Peck (McLaughlin, 1971). Therefore, it seems appropriate to generalize the results and apply them to most of the Homobasidiomycetes.

It is, furthermore, assumed that the main morphological features of mitosis and meiosis are basically identical. For example, if we are able to observe the intact nuclear envelope during mitosis in living material (Thielke, 1973, 1975), we would expect intact nuclear membranes in the thin sections of meiotic basidia. Cells prepared for electron microscopy, however, may contain artifacts.

Methods

The materials of the several species of Agaricales presented in this report were fixed in 2.5% glutaraldehyde, postfixed in 1% OsO_4, dehydrated in acetone, stained in 1% uranyl acetate dissolved in 70% acetone, and embedded in Spurr's (1969) plastic. Sections were prepared with a LKB ultramicrotome with glass knives, mounted on grids, and stained with 2% lead citrate. For electron microscopy they were examined in a Siemens Elmiskop 51 at 50 kV. The photographs were taken at a primary magnification of $\times$ 5000 (Figs. 3.1–3.3, 3.7–3.11, 3.14–3.19) or $\times$ 12,500 (Figs. 3.4–3.6, 3.12, 3.13).

The Sequence of Meiosis

Concerning the development within the hymenium it is necessary to distinguish between the asynchronous fungi (e.g., *Agaricus bisporus, Flammulina velutipes, Lentinus edodes,* and others) and the, relatively, synchronous ones of the genus *Coprinus.* In the species of the latter genus, however, there are differences in basidial ontogeny relative to the position within the pileus. The portions of the hymenia near the periphery develop earlier than those near the stipe. The basidia in the hymenia of many species of *Coprinus* are also dimorphic: i.e., there are shorter and longer basidia. The development of the shorter basidia is always retarded as compared to the development of the adjacent long basidia. The difference in time between both types of basidia within the same hymenium is sometimes as great as 1–2 h.

In both haploid and diploid nuclei, the nucleoli are always arranged excentrically near the inner membrane (Figs. 3.1, right basidium, 3.3, 3.6, 3.11, 3.19). In non-dividing nuclei an amorphous structure, the centriole-like organelle (centriole equivalent: CE; also termed spindle pole body: SPB) is attached to an indentation of the outer membrane of the nuclear envelope. Adjacent to the inner membrane, accumulated chromatin is visible radiating from several points of insertion (kinetochores?) into the inner part of the nucleus (Figs. 3.3, 3.4, 3.7, 3.18, 3.19). As far as we know, the dikaryon in young basidia occupies a central position within the cell. At the same site, fusion of these dikaryotic nuclei takes place. Very often, we have the impression that the centriole equivalents of the dikaryons initiate karyogamy (Thielke, 1976). As a result of the karyogamy the two nucleoli are present within the same envelope (Setliff et al., 1974; Thielke, 1974, 1976). Perhaps fusion of nucleoli initiates the synaptic structure at least for the nucleolus organizing chromosomes. The synaptic complexes develop during pachytene, as has been described by several other workers (Lu, 1966b; Volz et al., 1968; Thielke, 1971, 1974, 1976; Setliff et al., 1974; Gull and Newsam, 1975b). There does not seem to be any essential difference between these synaptonemal complexes and those of other organisms. The axis is inserted on the inner nuclear membrane and exhibits a very straight orientation (Figs. 3.6, 3.7). Sometimes it

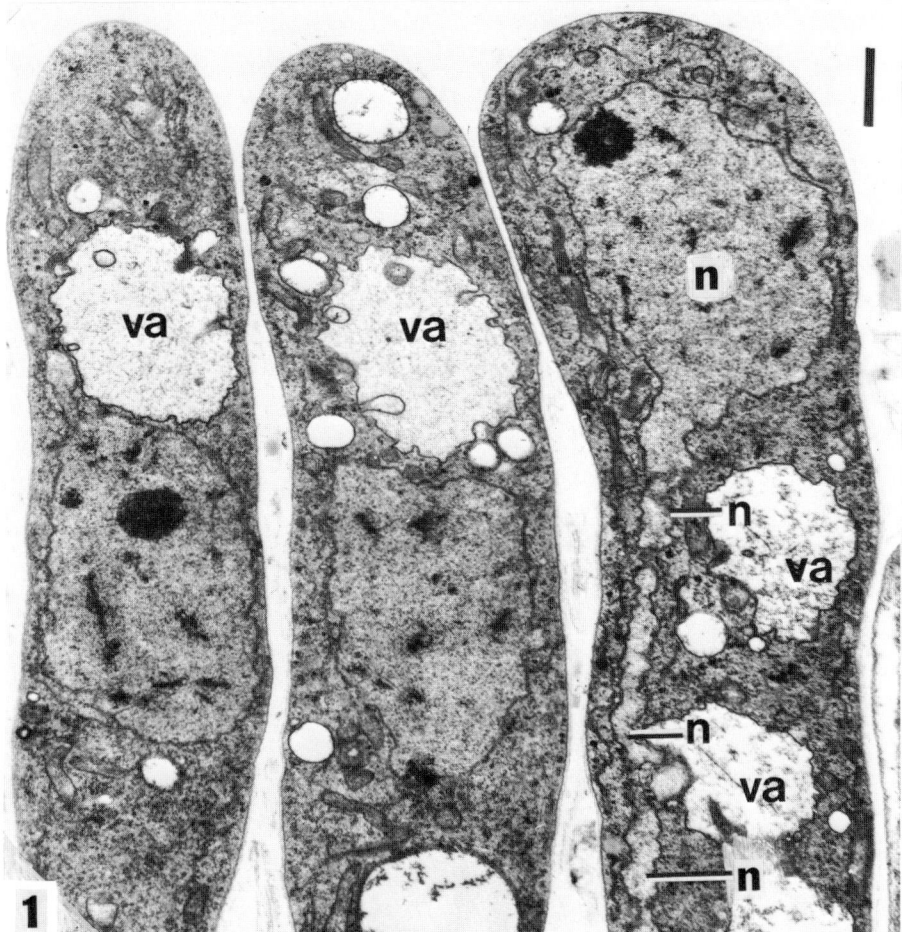

Fig. 3.1. Premeiotic basidia in the hymenium of *Lentinus edodes*. In the right basidium the nucleus (n) is migrating to the apex of the basidium leaving a tail-like projection; va, = vacuole; bar = 1 μm.

appears that both ends of the axis are fastened to the inner surface of the nuclear envelope (Fig. 3.8).

Most of the young basidia develop an obvious vacuole near the apex. Just before meiosis, the large diploid nucleus migrates to the apex of the basidim and passes this vacuole (Wells, 1965). This locomotion seems to cause a deformation of the nucleus and leaves a tail-like trace because of its flexibility (Fig. 3.1, right basidium).

Nuclear division occurs at the apex of the basidia (Fig. 3.2). In species of *Coprinus* the chances of finding dividing nuclei in adjacent basidia of the same length are good. To determine whether there is an intranuclear spindle or whether polar gaps exist, it should be taken into consideration that fixation artefacts are

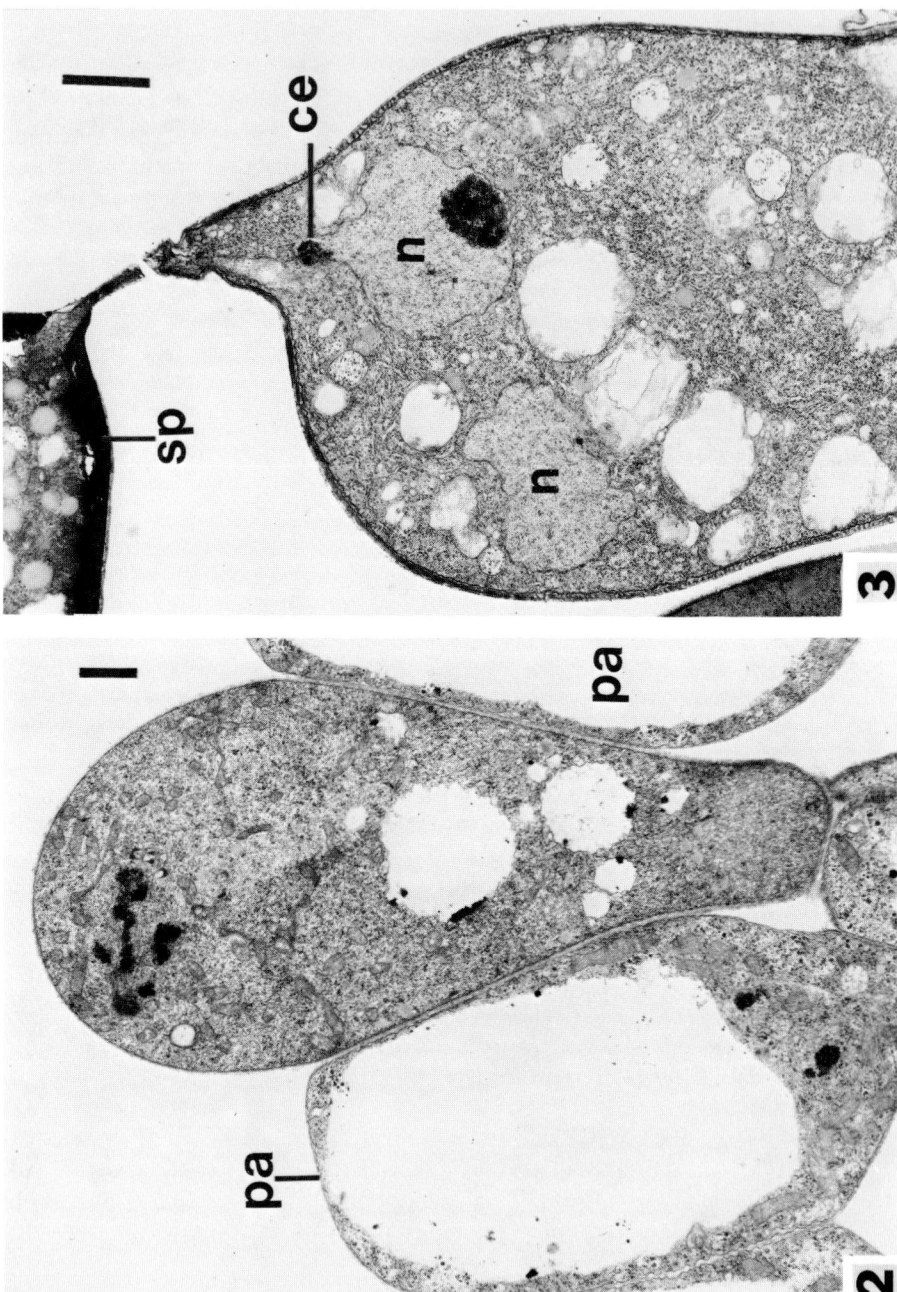

possible. If we establish the presence of a membrane in a fixed section, we believe there is no doubt that this membrane must also be really present in the living cell. If we notice an interrupted membrane in a fixed section, it is possible that this membrane was damaged during fixation or the membrane might be out of the plane of the section. Because the spindle pole is much smaller than the rest of the nucleus it is not always situated in a favorable plane of the particular section. From this region of the nucleus, we often get oblique sections. Moreover, it should be considered that biological membranes behave in a flexible manner, as does the surface of mercury; they can proliferate and contract within a very short time. We know from the flow of membranes that they may contact and join with each other, and that they might be interrupted. Of course these effects could also be caused by the chemicals used in fixation.

From the situation demonstrated in Fig. 3.5, it is obvious that the tonoplast has broken and has partially left its original position. Because of the flexibility of this membrane, a row of vesicles marks the former border to the cytoplasm, whereas the retracting parts of the same tonoplast contact each other to form a new membrane beyond the earlier location. Perhaps due to osmotic differences the mesosome-like body has been created as an artefact. Analogous artefacts seem to have been derived from the plasmalemma in Fig. 3.4. Following such observations the question arises as to the real nature of the "lomasomes" and "plasmalemmasomes" (Marchant and Moore, 1973).

At the onset of the first meiotic division, both globular ends of the centriole-like organelle separate, and microtubules radiate from one pole to the other. At this stage, the electron-dense middle piece has vanished or its structure has changed. The polar microtubules are surrounded by the chromatin that seems to have an obvious affinity to the nuclear envelope (Fig. 3.10). In another plane microtubules are running from one pole to the kinetochores of the chromatin. The entire spindle apparatus is formed across the apex of the basidia. In the meantime, the nucleolus has vanished as indicated by the free ribosomes remaining within the nucleus. Both polar bodies are now surrounded by the nuclear membrane; they must have been invaginated into the nucleoplasm.

Evidence from fluorescence microscopy has created the impression that at this stage the chromatin is assembled in a ring or cylinder-like structure. Perhaps this ring consists of 4 particles, as was suggested by Setliff et al. (1974) for *Poria latemarginata*. In sections prepared for electron microscopy, we have to imagine that this cylinder is crossed by the axis of the spindle. Favorable longitudinal sections demonstrate that the axis of the spindle consists of several microtubules. It is impossible to identify a stage corresponding to metaphase and a configuration corresponding to a metaphase plate. Two different aspects of the structure of the chromatin are evident during both meiotic divisions. In the plane of the spindle axis the strand of microtubules is flanked by the chromatin mass (Fig. 3.10).

◀ **Figs. 3.2–3.3.** A short basidium of *Coprinus cinereus* with a dividing nucleus between two paraphyses (pa); bar = 1 µm. Fig. 3.3 A postmeiotic basidium of *Stropharia rugosoannulata* with two nuclei (n). The right nucleus is moving towards the sterigma into the spore (sp) with the centriole equivalent (ce) at the leading edge; bar = 1 µm.

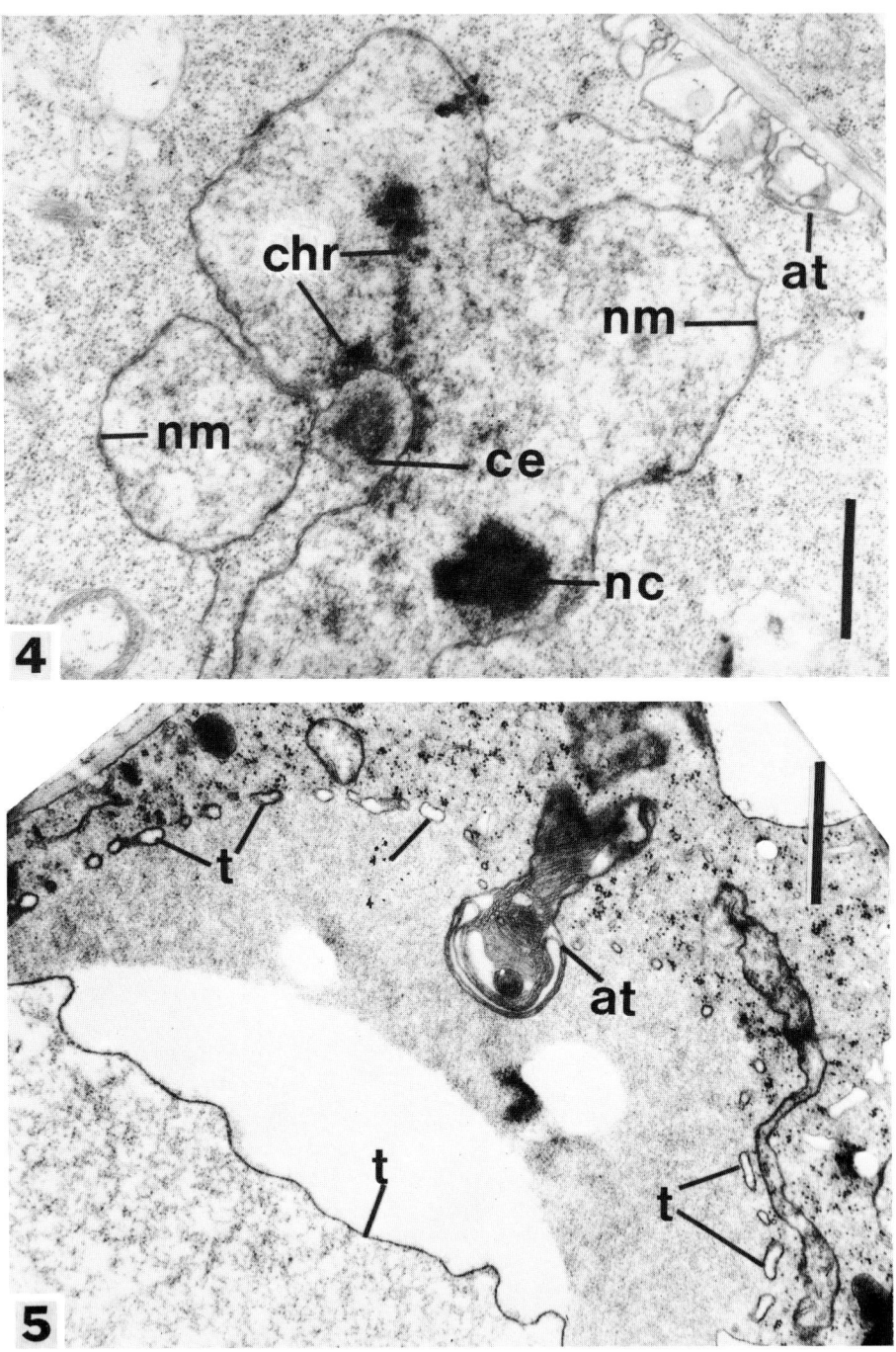

Figs. 3.4–3.5. A tangential section of the surface of the diploid nucleus of *Coprinus cinereus*. The chromatin (chr) is inserted at the nuclear membrane (nm) near the centriole equivalent (ce) which is outside the envelope. The lomasome-like structure (at) is considered to be an artefact; bar = 1 μm. Fig. 3.5 A vacuole of a premeiotic basidium of *Flammulina velutipes* demonstrating the effects of bad fixation. The tonoplast has withdrawn from a region represented by a row of vesicles. A membranous structure (at), perhaps derived from the tonoplast, is presumed to be an artefact; bar = 1 μm.

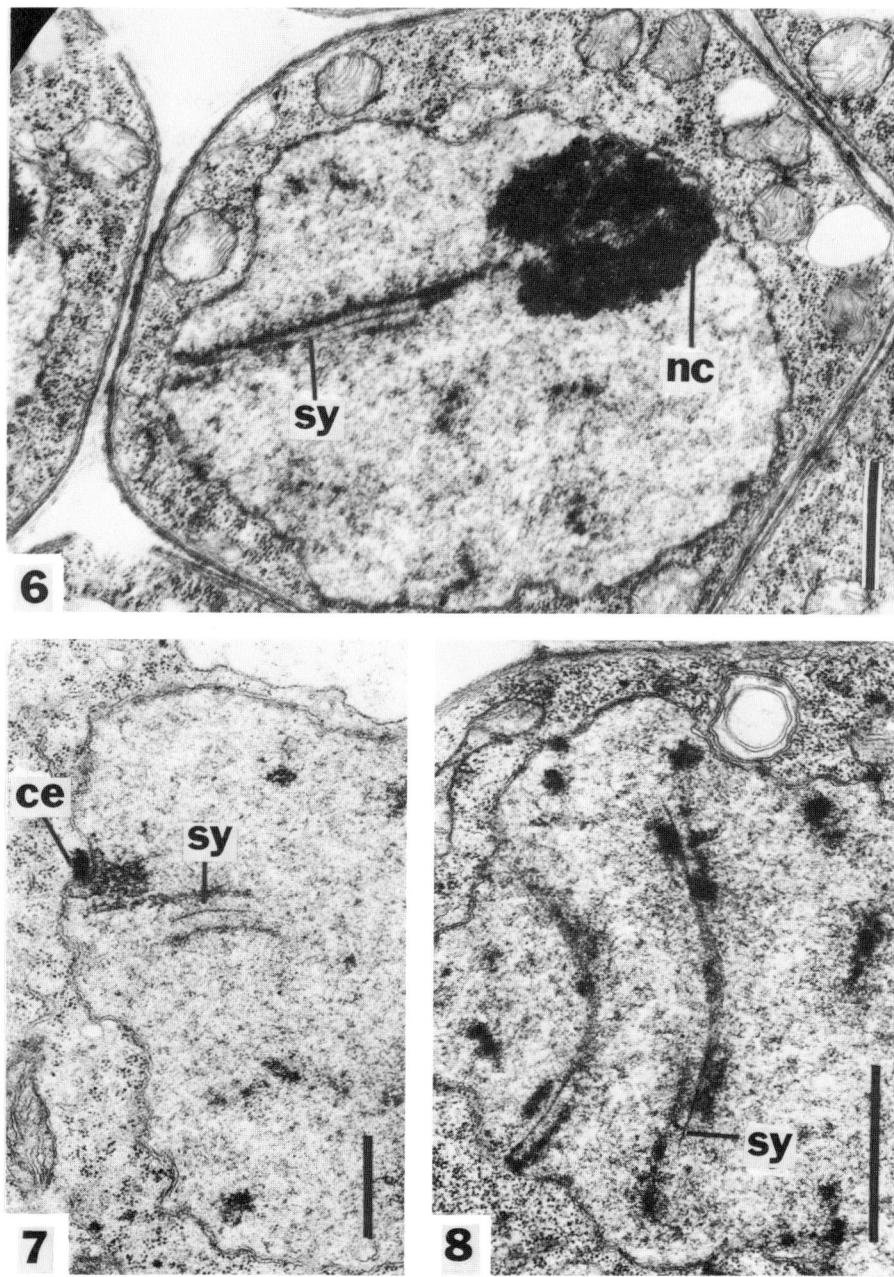

Figs. 3.6–3.8. A synaptonemal complex (sy) in a basidium of *Psilocybe turficola* that connects the nucleolus (nc) and the nuclear envelope; bar = 1 μm. Fig. 3.7 A synaptonemal complex (sy) inside the nucleus of *Lentinus edodes* near the centriole equivalent that is attached to the outer nuclear membrane; bar = 1 μm. Fig. 3.8 Synaptonemal complexes (sy) in *Psilocybe turficola* that seem to be attached to the inner membrane of the nuclear envelope; bar = 1 μm.

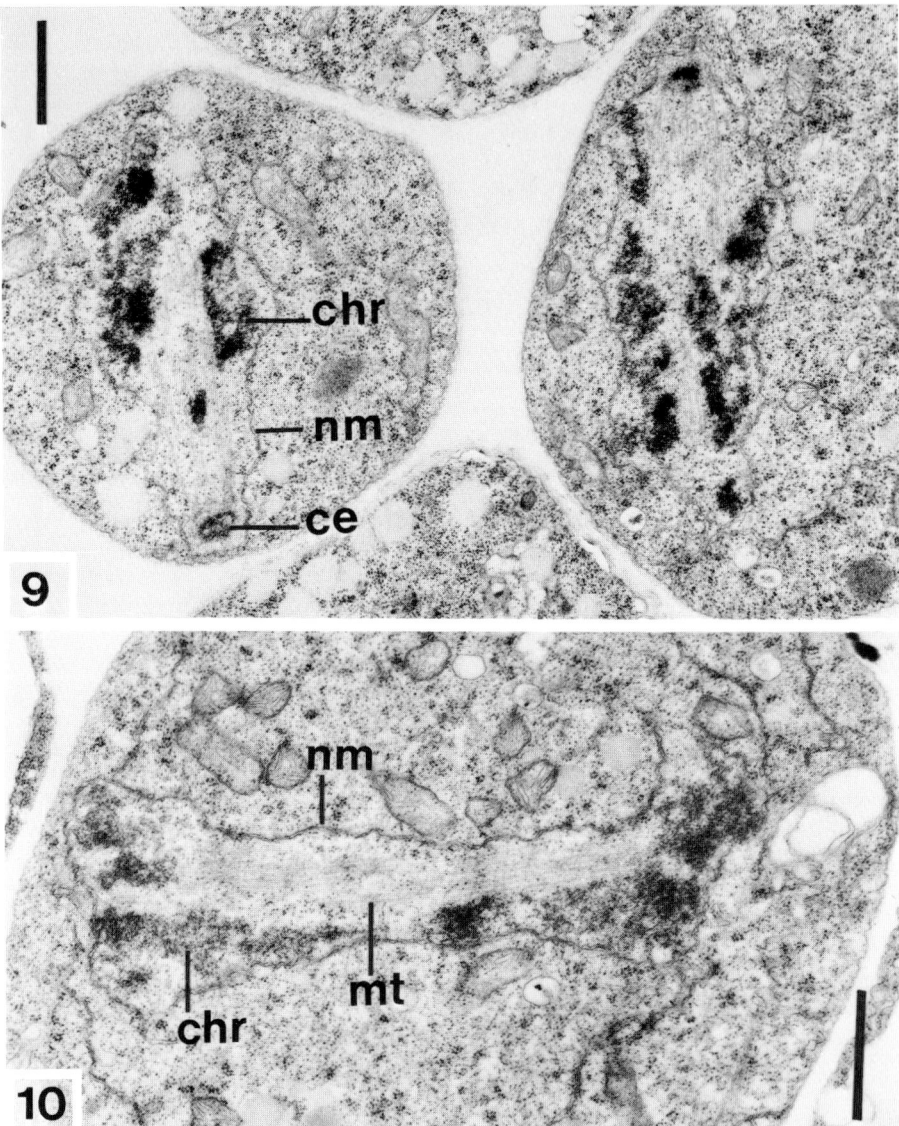

Figs. 3.9–3.10. Cross-section through the hymenium of *Coprinus micaceus* demonstrating intranuclear meiotic divisions in 2 adjacent basidia. The plane of the section is median to the lower spindle pole body (ce) in the left basidium; chr, chromatin; nm, nuclear membrane; bar = 1 μm. Fig. 3.10 First meiotic division within an intact nuclear membrane (nm) in *Coprinus micaceus*. Polar microtubules (mt) and chromatin (chr) are at a stage that is assumed to be anaphase. The section is not a median section; bar = 1 μm.

Within the same plane of a median section we can either observe no chromosomal microtubules or kinetochores, or only a few such structures are visible (Figs. 3.15, 3.16). The second aspect consists predominately of chromosomal microtubules in non-median sections of the spindle apparatus (Figs. 3.13, 3.14). Sometimes we can distinguish more than one kinetochore in connection with the same chromosomal complex (Fig. 3.13). Very often a stage of separation of homologous material can be identified. Because this separation of homologous material does not occur symmetrically, this process was considered to have an asynchronous disjunction. If one imagines that the chromosomal cylinder has a wavy surface, however, this phenomenon can be explained by spatial differences instead of temporal ones. The fact that in median sections both aspects exclude each other confirms the assumption that there is a metaphase cylinder instead of a metaphase plate.

Late anaphase is characterized by the dumbbell shape of the dividing nucleus that contains the polar bodies with elongated polar microtubules flanked by the chromatin inside the intact nuclear membrane. During late telophase the former spindle poles are temporarily enclosed by the constricting nuclear membrane. At the end of this phase the spindle poles are totally evaginated, as was initially demonstrated in meiotic division in *Coprinus radiatus* and *Agaricus bisporus* (Thielke, 1971, 1974, 1976). During meiotic interphase both nuclei are again situated in the middle region of the basidium, and the centriole equivalents are again in their diglobular form (Fig. 3.11) with an electron-dense isthmus connecting the amorphous globular poles. At the onset of the second division the nuclei return to

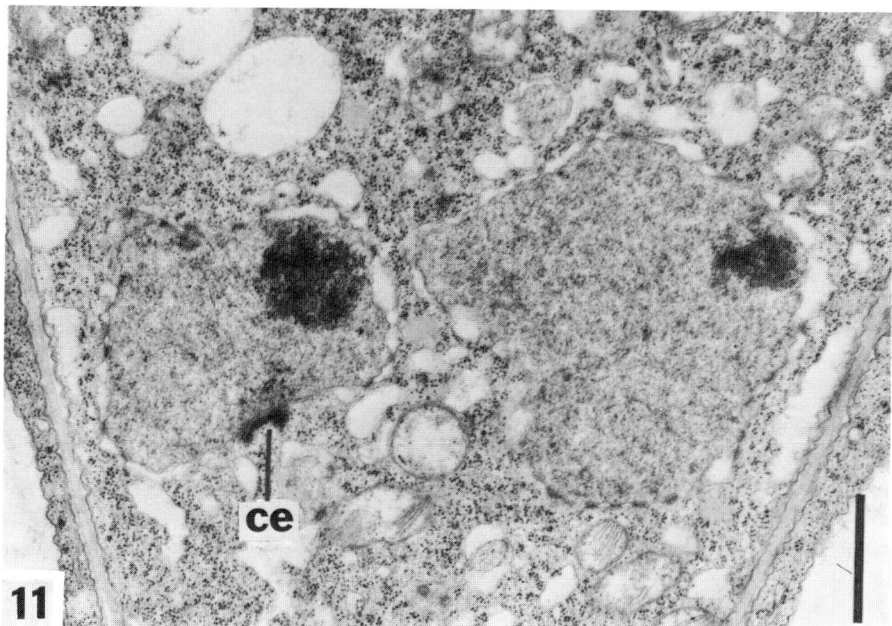

Fig. 3.11. Two interphase nuclei in a basidium of *Coprinus cinereus* after the first meiotic division. The diglobular centriole equivalents (ce) are outside the nuclear membrane; bar = 1 μm.

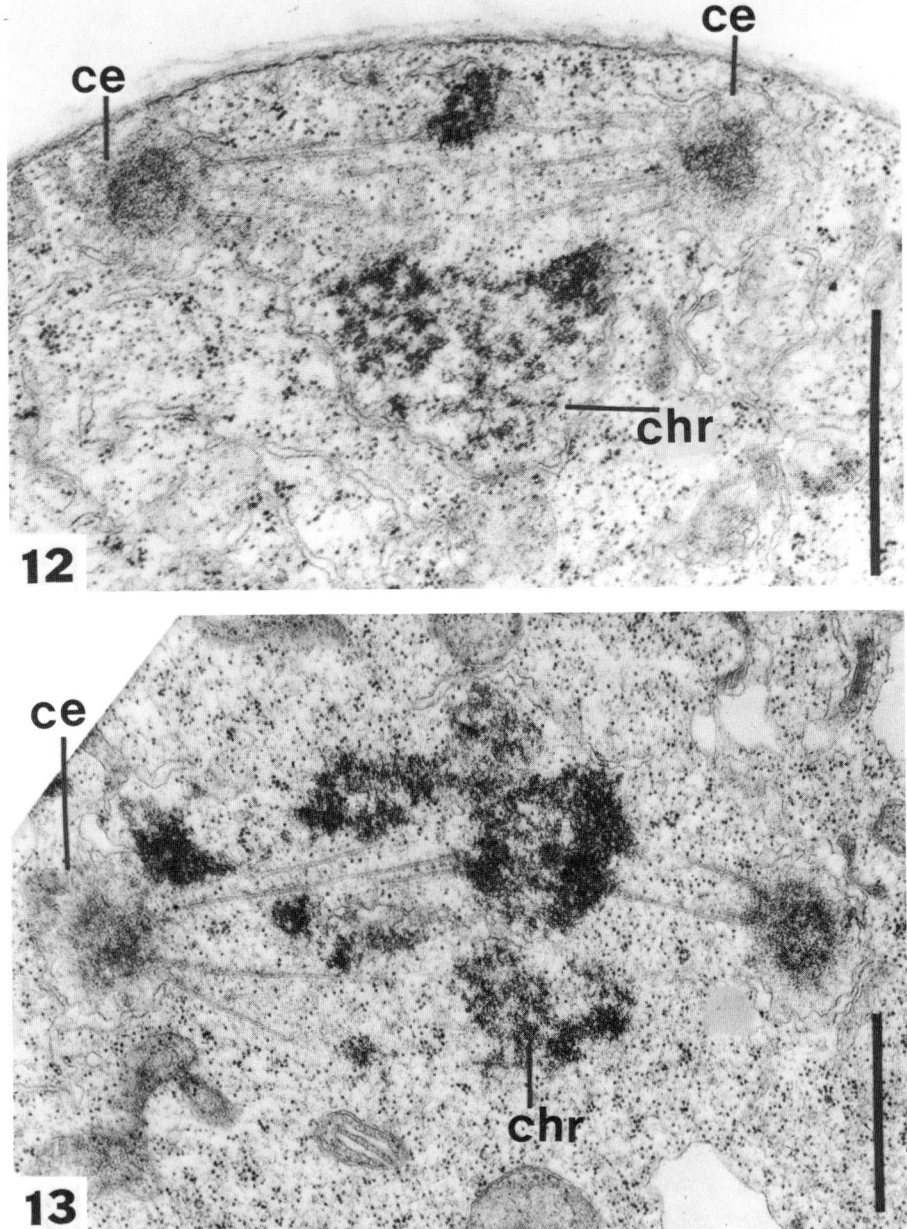

Figs. 3.12–3.13. Polar and chromosomal microtubules between the spindle poles (ce) during the second meiotic division in *Coprinus cinereus*. The nuclear membrane is intact; chr, chromatin; bar = 1 μm. Fig. 3.13 Chromosomal microtubules during the second meiotic division in *Coprinus cinereus* at a stage that seems to correspond to metaphase; ce, centriole equivalent; chr, chromatin; bar = 1 μm.

an apical position in the basidium, where they divide in a manner that resembles the first process as far as the behavior of the spindle apparatus is concerned. At the end of the second division the evagination of the centriole equivalent is repeated (Figs. 3.17–3.19). Because this evagination always occurs near the periphery of the cell, the outer membrane of the constricted nuclear envelope comes into contact with the plasmalemma and seems to be absorbed there. In addition, the former polar body undergoes some sort of morphogenesis: a dense particle separates from an amorphous particle and vanishes near the plasmalemma (Fig. 3.17). Therefore, at the end of the second telophase the 4 centriole equivalents are regularly oriented towards the cell periphery (Figs. 3.18, 3.19).

On the surface of these postmeiotic nuclei, the diglobular form of the centriole equivalent has not been recognized. In the Agaricales mentioned above a postmeiotic mitosis was not observed. The telophase basidia are temporarily characterized by a special cytoplasmic pattern. The cytoplasm becomes denser, more ribosomes appear in the form of polysomes, and the ER and the nuclear envelope are somewhat inflated. These changes may be reactions due to the fixing procedure because of the special sensitivity of the cytoplasm at this stage. This seems to be the case at the end of the first as well as at the end of the second division (Figs. 3.11, 3.18).

When the postmeiotic basidia begin to develop spores, as described by McLaughlin (1977) for *Coprinus cinereus,* each nucleus moves through a sterigma. In this case, too, the centriole-like organelle acts as a moving center. It regularly precedes the migrating nucleus (Fig. 3.3), from which a bundle of cytoplasmic microtubules emanate, through the sterigma, into the cytoplasm of the spore (Thielke, 1978).

Discussion

The Behavior of the Nuclear Membrane

In contrast to the ascomycetes, nuclear divisions in the basidiomycetes are said to possess polar gaps or fenestrae during both late anaphase and telophase (Fuller, 1976; Wells, 1977). Based on my own studies of mitosis in living material, I think that the morphological behavior of the nuclear envelope during both mitotic and meiotic divisions is essentially identical in all groups of the higher fungi. The data given in this report seem to support this assumption. During my own studies on meiotic basidia in *Coprinus radiatus* it was found that the completeness of a membrane depended on the fixing procedure. Lerbs (1971) reported a fragmentation of the envelope during meiosis in *C. radiatus*. Using the same strain of *C. radiatus* but a different fixing method I was able to demonstrate that the nuclear envelope remains intact during the same stages that Lerbs had reported a fragmentation. It is assumed, therefore, that the polar gaps developed during fixation, perhaps from nuclear pores, or that the membrane was interrupted in the way described in the tonoplast in Fig. 3.5.

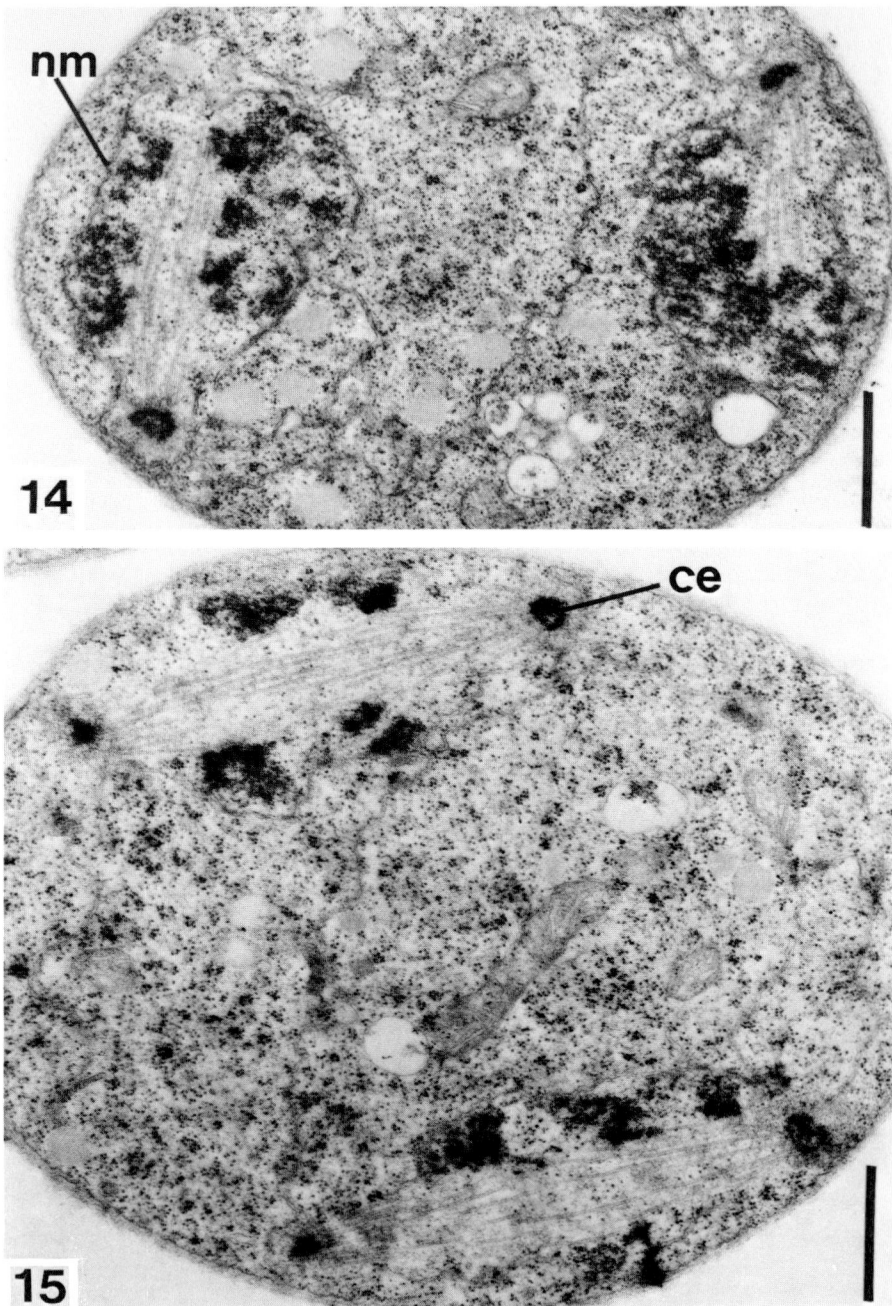

Figs. 3.14–3.15. An asymmetric section through a basidium of *Coprinus micaceus* during the second division of meiosis; nm, nuclear membrane; bar = 1 μm. Fig. 3.15 Polar microtubules surrounded by chromatin during the second meiotic division in *Coprinus micaceus;* ce, centriole equivalent; bar = 1 μm.

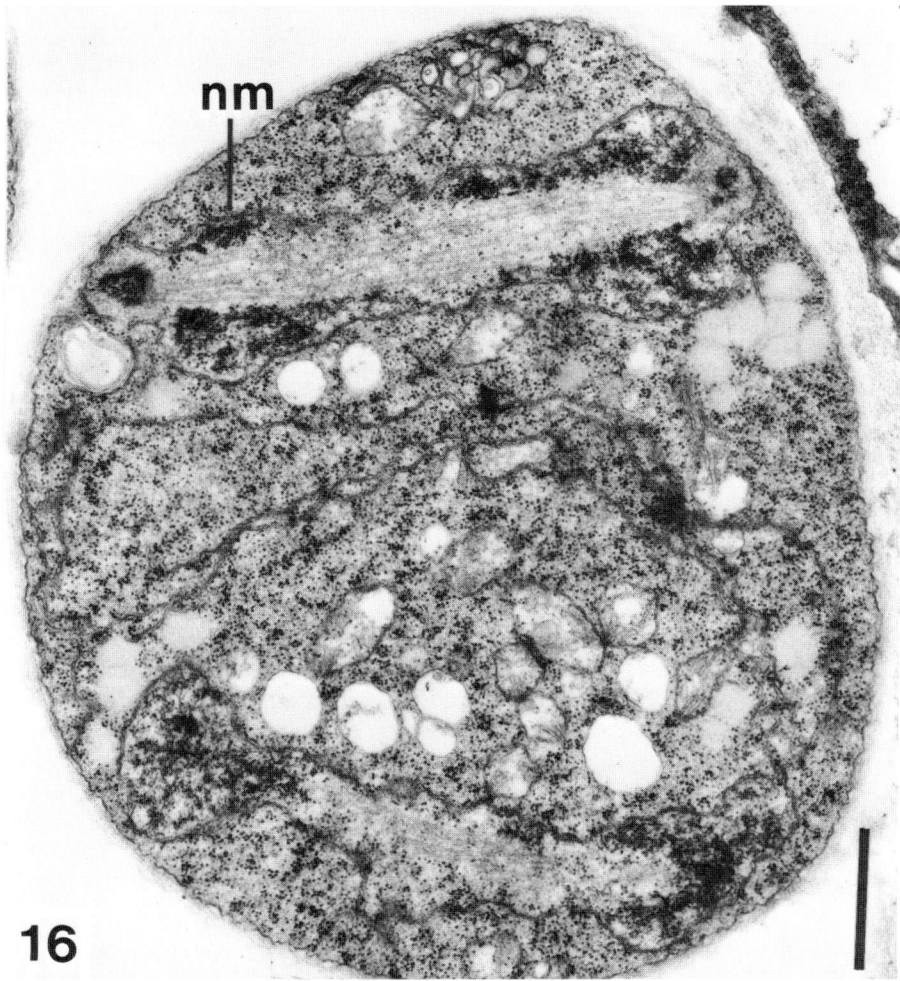

Fig. 3.16. Second meiotic division in *Coprinus micaceus*. In the upper figure polar microtubules are visible within an intact nuclear envelope (nm); bar = 1 μm.

The special flexibility of the nuclear membrane is obvious when observing the flow of this membrane from the surface of the nucleus to the surface of the centriole equivalent and later on to the plasmalemma. This flow demonstrates clearly that the inner as well as the former outer membrane are capable of joining with the plasmalemma. In species in which nuclear divisions occur rapidly the absorption of membranes by the plasmalemma does not keep pace with their production during nuclear division. Therefore, an accumulation of membranes derived from the nuclear membranes occurs as was shown by Lerbs (1971) and Thielke (1974). Such clusters have recently been observed in *Coprinus cinereus* and *C. micaceus*. Occasionally it has been suggested that aggregates of vesicles near the nucleus, described as dictyosomes (Moore, 1963; Lu, 1966b), were derived from these membranes.

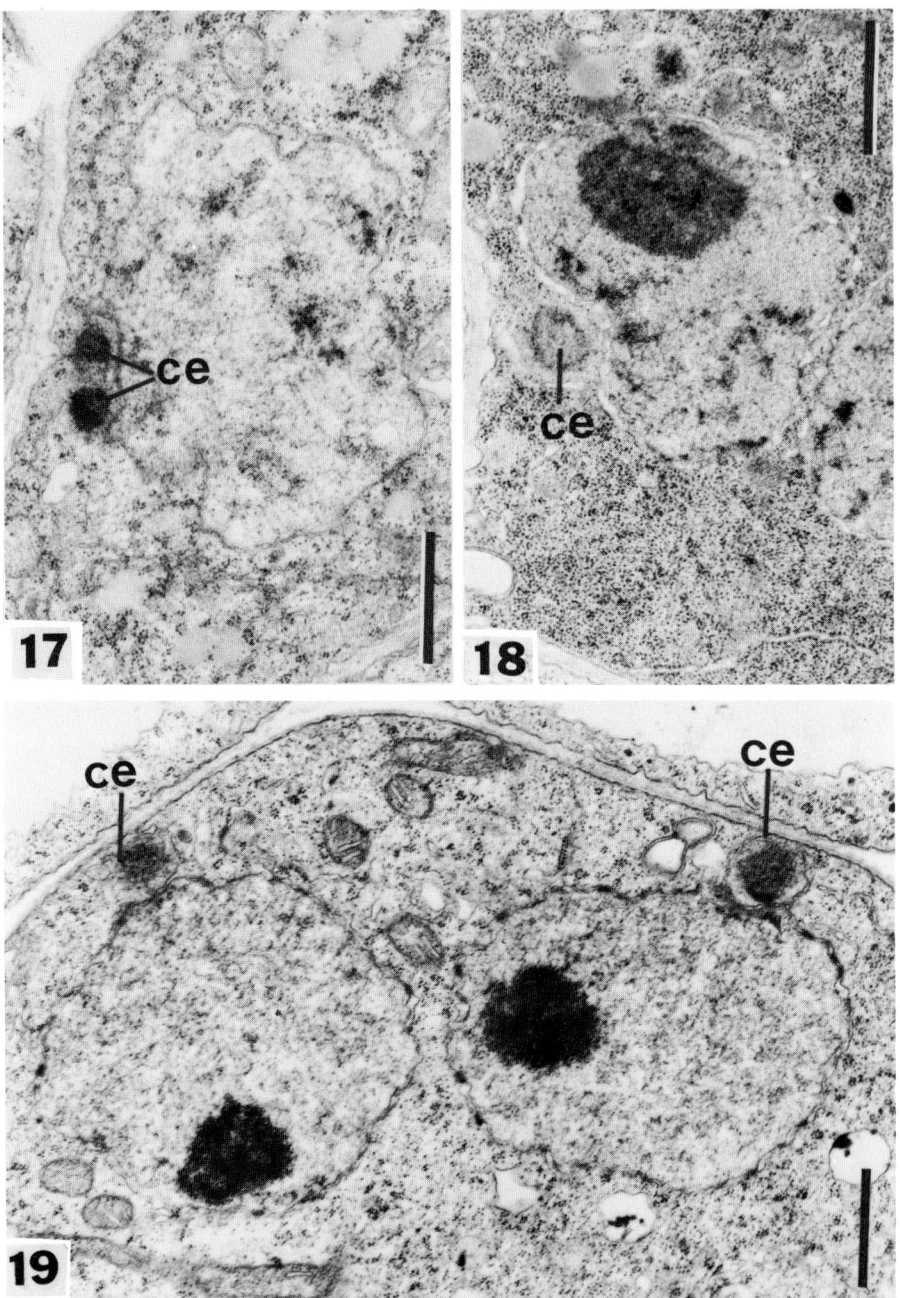

Figs. 3.17–3.19. Postmeiotic nucleus in *Coprinus micaceus* with a centriole-like organelle (ce) consisting of two particles outside the nuclear membrane; bar = 1 μm. Fig. 3.18 A postmeiotic nucleus in *Coprinus cinereus* with a centriole equivalent (ce) that is partially surrounded by the nuclear envelope. The cytoplasm is very dense and contains many ribosomes; bar = 1 μm. Fig. 3.19 Two of the 4 postmeiotic nuclei in a basidium of *Coprinus cinereus*. The centriole equivalents (ce) are partially surrounded by the nuclear envelope and are located near the plasmalemma; bar = 1 μm.

Morphology of Chromosomes

Evidence from very careful light-microscopic studies seems to support the concept of the presence of very small individual chromosomes. Valuable data were provided by Olive (1953, 1965). However, this material was first fixed and stained by techniques that may have resulted in artefacts. Evidence from vital staining may give us another interpretation. Fluorescent dyes are very useful as they can be used at very low concentrations. Moreover, it is possible to determine whether the cell is still viable. Following the application of the dye Coriphosphin (Keeble and Jay, 1962), it is possible to distinguish between the RNA of the nucleolus and the DNA of the chromatin. It has been observed that, when the nucleolus vanishes during early prophase, condensation of chromatin will occur near the nuclear membrane. This chromosomal material seems to behave according to a modified two-track model of nuclear division as proposed by Day (1972) and was demonstrated for mitosis in a species of *Puccinia* (Harder, 1976). During an advanced stage of prophase, a view at right angles to the long axis of the spindle illustrated the ring-like arrangement of the chromatin.

Electron-microscopic research suggests a special affinity of the chromatin for the inner surface of the nuclear envelope. We, therefore, should perhaps consider the nuclear membrane as being actively engaged in the early process of chromosome separation. Reports about "parallel rows of chromatin" (Robinow and Caten, 1969; Day, 1972) might be explained by the fact that the plane of the section was parallel with the long axis of the spindle.

The Kinetic Apparatus

The polar region of a dividing nucleus is represented by an amorphous structure that has been designated with a variety of terms. The structure and function of this organelle has been described in detail (Girbardt, 1971; McLaughlin, 1971; Raju and Lu, 1973; Setliff et al., 1974; Girbardt and Hädrich, 1975; Gull and Newsam, 1975a) and reviewed by Kubai (1975), Fuller (1976), and Wells (1977). At the present time the term "spindle pole body" (SPB) has generally been accepted. This term was proposed when the behavior during interphase was unknown. In comparison to a true centriole, this organelle does not possess the characteristic pinwheel structure, as is also the case in all organisms in which flagellae do not develop. The function, however, of this polar body as compared to the function of a true centriole should fulfill the criteria described by Fulton (1971): (1) In non-dividing nuclei this body marks the position of the chromatin that is attached to the inner membrane of the nuclear envelope. (2) The presence of this organelle provides a basis for the development of a new one. (3) It functions as a microtubule organizing center (MTOC) as discussed by Pickett-Heaps (1969), and in this connection is responsible for the spindle as well as for the locomotion of the nucleus.

From this point of view, it is appropriate to consider the SPB as an organelle homologous to a true centriole. Since the function of developing the microtubules of the flagellae has become lost during the evolution of higher fungi, the pinwheel structure has become useless. From my experience this polar body is in contact with the nuclear membrane throughout its existence. The diglobular form is present merely in an extranuclear position during interphase when the chromatin is inserted near the middle piece at the opposite region of the nuclear envelope. At the onset of the nuclear division, only the monoglobular form occurs inside the nuclear membrane. At this time the electron-dense middle piece has vanished. Therefore, it seems logical to suggest that the isthmus of the diglobular form might represent an inactive stage of the kinetochores. This might be the reason why in higher fungi, as well as in lower ones, the chromatin is attached on the inner membrane of the nuclear envelope adjacent to the polar body. This unusual attachment of the chromatin may be involved in the special behavior of chromosomal material and of nuclear membranes during nuclear division in the higher fungi.

Acknowledgments. I wish to thank Mrs. Arnhild Lederer for her valuable assistance in preparing the material for electron microscopy and for the micrographs in Figs. 3.5–3.8, 3.12, 3.17, and 3.18.

References

Day, A. W.: Genetic implications of current models of somatic nuclear division in fungi. Can. J. Bot *50*, 1337–1347 (1972).
Esser, K., Kuenen, R.: Genetik der Pilze. Berlin: Springer-Verlag 1967.
Fuller, M. S.: Mitosis in fungi. Int. Rev. Cytol. *44*, 113–153 (1976).
Fulton, C.: Centrioles. In: Results and Problems in Cell Differentiation. Vol. II. Origin and continuity of cell organelles. Reinert, J., Ursprung, H. (eds). Berlin: Springer-Verlag 1971, pp. 170–221.
Girbardt, M.: Ultrastructure of the fungal nucleus. II. The kinetochore equivalent (KCE). J. Cell Sci. *2*, 453–473 (1971).
Girbardt, M., Hädrich, H.: Ultrastruktur des Pilzkerns. III. Genese des Kern-assoziierten Organells (NAO = "KCE"). Z. Allg. Mikrobiol. *15*, 157–173 (1975).
Gull, K., Newsam, R. J.: Meiosis in basidiomycetous fungi. I. Fine structure of spindle pole organization. Protoplasma *83*, 247–257 (1975a).
Gull, K., Newsam, R. J.: Meiosis in basidiomycetous fungi. II. Fine structure of the synaptic complex. Protoplasma *83*, 259–268 (1975b).
Gull, K., Newsam, R. J.: Meiosis in the basidiomycetous fungus *Coprinus atramentarius*. Protoplasma *90*, 343–352 (1976).
Harder, D. E.: Mitosis and cell division in some cereal rust fungi. II. The process of mitosis and cytokinesis. Can. J. Bot. *54*, 995–1019 (1976).
Keeble, S. A., Jay, R. F.: Fluorescent staining for the differentiation of intracellular ribonucleic acid and desoxyribonucleic acid. Nature (Lond.) *193*, 695–696 (1962).
Kubai, D. F.: The evolution of the mitotic spindle. Int. Rev. Cytol. *43*, 167–227 (1975).
Lerbs, V.: Licht- und elektronenmikroskopische Untersuchungen an meiotischen Basidien von *Coprinus radiatus* (Bolt) Fr. Arch. Mikrobiol. *77*, 308–330 (1971).

Lu, B. C.: Golgi apparatus of the Basidiomycete *Coprinus lagopus.* J. Bacteriol. *92,* 1831–1834 (1966a).
Lu, B. C.: Fine structure of meiotic chromosomes of the basidiomycete *Coprinus lagopus.* Exp. Cell Res. *43,* 224–227 (1966b).
Lu, B. C.: Meiosis in *Coprinus lagopus:* A comparative study with light and electron microscopy. J. Cell Sci. *2,* 529–536 (1967).
Marchant, R., Moore, R. T.: Lomasomes and plasmalemmasomes in fungi. Protoplasma *76,* 235–247 (1973).
McLaughlin, D. J.: Centrosomes and microtubules during meiosis in the mushroom *Boletus rubinellus.* J. Cell Biol. *50,* 737–745 (1971).
McLaughlin, D. J.: Basidiospore initiation and early development in *Coprinus cinereus.* Am. J. Bot. *64,* 1–16 (1977).
Moore, R. T.: Fine structure of Mycota. II. Occurence of the golgi dictyosome in the heterobasidiomycete *Puccinia podophylli.* J. Bacteriol. *86,* 866–871 (1963).
Olive, L. S.: The structure and behavior of fungus nuclei. Bot. Rev. *19,* 439–586 (1953).
Olive, L. S.: Nuclear behavior during meiosis. In: The Fungi. Vol. I. Ainsworth, G. C., Sussman, A. S. (eds). New York: Academic Press 1965, pp. 143–161.
Pickett-Heaps, J. D.: The evolution of the mitotic apparatus. An attempt of comparative ultrastructural cytology in dividing plant cells. Cytobios *3,* 257–280 (1969).
Raju, N. B., Lu, B. C.: Meiosis in *Coprinus.* IV. Morphology and behavior of spindle pole bodies. J. Cell Sci. *12,* 131–141 (1973).
Robinow, C. F., Caten, C. E.: Mitosis in *Aspergillus nidulans.* J. Cell Sci. *5,* 403–431 (1969).
Setliff, E. C., Hoch, H. C., Patton, R. F.: Studies on nuclear division in basidia of *Poria latemarginata.* Can. J. Bot. *52,* 2323–2333 (1974).
Spurr, A. R.: A low-viscosity epoxy resin embedding medium for electron microscopy. J. Ultrastruct. Res. *26,* 31–43 (1969).
Thielke, C.: New investigations on the structure of mushrooms. IX. Int. Congr. Mushr. Science, 285–293 (1971).
Thielke, C.: Intranucleäre Mitosen in homokaryotischen und dikaryotischen Mycelien der Basidiomyceten. Arch. Mikrobiol. *94,* 341–350 (1973).
Thielke, C.: Intranucleäre Spindel und Reduktion des Kernvolumens bei der Meiose von *Coprinus radiatus* (Bolt.) Fr. Arch. Microbiol. *98,* 225–237 (1974).
Thielke, C.: *Polyporus adustus* (Polyporaceae) Intranucleäre Mitose im dikaryotischen Mycel. Göttingen: Encyclopaedia Cinematographica 1975, pp. 1–9.
Thielke, C.: Intranucleäre Meiose bei *Agaricus bisporus.* Z. Pilzkd. *42,* 57–66 (1976).
Thielke, C.: Feinstrukturen bei Basidiomyceten. Z. Mykol. *44,* 71–89 (1978).
Volz, P. A., Heintz, C. H., Jersild, R., Niederpruem, D. J.: Synaptinemal complexes in *Schizophyllum commune.* J. Bacteriol. *95,* 1476–1477 (1968).
Wells, K.: Ultrastructural features of developing and mature basidia and basidiospores of *Schizophyllum commune.* Mycologia *57,* 236–261 (1965).
Wells, K.: Meiotic and mitotic divisions in the Basidiomycotina. In: Mechanisms and Control of Cell Division. Rost, T. L., Gifford, E. M., Jr. (eds.). Strandsburg, Pa.: Dowden Hutchinson & Ross 1977, pp. 337–374.

Chapter 4

Replication of Deoxyribonucleic Acid and Crossing Over in *Coprinus*

BENJAMIN C. LU

Introduction

Coprinus cinereus (Schaeff. ex Fr.) S. F. Gray (= *C. lagopus* sensu Buller) is a member of the Agaricales. The species of *Coprinus* are probably unique in having a naturally evolved synchronous meiotic system. In a single fruiting body of *C. cinereus*, there are about 100 gills, all of which are at the same stage of development. There are approximately $1-3 \times 10^7$ basidia in a fruiting body, of which about 70–75% are at the same stage of meiosis. The lower part of the gill is usually slightly more advanced in development than that of the apex but the difference in time is within a 2-h period (Raju and Lu, 1971).

The advantages of such a synchronous meiotic system need no elaboration. Several types of research have been simplified in this system, namely studies of the meiotic cell cycle, DNA replication, and genetic recombination. In this presentation, I shall review some of the pertinent data on DNA synthesis and genetic recombination in the light of more recent discoveries.

An Overview of the System

The nature of a gill fungus provides a convenient means of monitoring the progress in development. A few gills can be removed from a fruiting body for cytological examination without damaging the continued normal progress of the rest of the gills. Since all gills are identical in development, a time course study of meiotic processes can readily be achieved (Raju and Lu, 1971). The meiotic cell cycle and some important events are summarized in Fig. 4.1.

At the beginning of the S phase, two events occur: (1) initiation of DNA synthesis and (2) the DNA chain elongation. The first step can be arrested by a shift-up to high temperature (35°C) and a continuous light regime. As shown in Fig. 4.2, either condition alone is ineffective, but the two combined (to be called the restrictive condition) prevent the initiation of DNA synthesis (Lu, 1972, 1974a; Lu and Jeng, 1975). Thus, by incubating fruiting cultures in the restrictive condition, all basidia can be held at the pre–S phase stage. Upon returning to the normal culture temperature at 25°C, all basidia enter S phase immediately. This

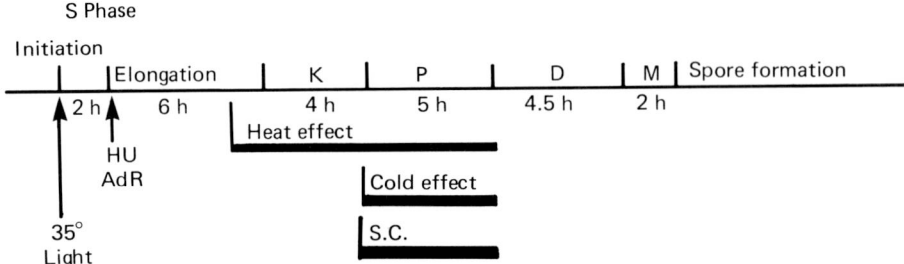

Fig. 4.1. A summary of the meiotic events and their timing in *Coprinus cinereus*. The S phase involves the initiation of DNA replication and the DNA chain elongation. Only the initiation step is subject to the arrest by high temperature and continuous light regime, but the chain elongation is sensitive to inhibitors such as hydroxyurea (HU) and 2-deoxyadenosine (AdR). K, karyogamy; P, pachytene; D, diplotene; M, meiotic divisions; S.C., the synaptonemal complex.

arrest–release technique can therefore be used to control the initiation of meiosis (Lu, 1974a).

The general culturing procedure is as follows: Cultures are prepared on horse dung or on agar medium and inoculated with dikaryotic mycelium (A_5B_5 *den* × A_2B_3 *met−1*) on day-1. These cultures are incubated at 35°C in absolute darkness for 5 days, at which time the mycelium has completely covered the surface of the medium. On day-6 at 1600 hours (clock time) when the light cycle begins, the darkgrown cultures were photoinduced to fruit in an incubator at 25°C with light regime of 16 h light–8 h darkness. This time regime was chosen so that karyogamy in normal cultures usually begins at 0900–1000 hours (clock time) on day-11 under our laboratory conditions. I say "usually" because karyogamy has occasionally been found to vary a day in either direction; temperature, cultural conditions, age of inoculum can all influence the timing of development.

To overcome this handicap, cytological monitoring is used. A few gills are removed from a fruiting body, fixed for 20 sec in a drop of Carnoy solution (Lu, unpublished). A small piece of the gill tissue is removed from the midsection of the gill for staining with iron haematoxylin. The whole process takes only 1–2 min. The nucleoli are stained darkly, thus prekaryogamy and postkaryogamy basidia can be clearly distinguished because the former has two nucleoli and the latter has only one. The percentage of postkaryogamy basidia indicates the process of karyogamy. The onset of karyogamy is the first important landmark of meiosis as it can be unmistakably identified. Once the onset of karyogamy is determined, the processes can be reasonably timed according to the scheme shown in Fig. 4.1. The second landmark in meiosis is the onset of metaphase I. Again it can be easily identified by cytological monitoring. Both landmarks serve as reference points to place meiotic events accurately.

To control meiosis, cultures are shifted to the restrictive condition on day-10, a few hours prior to the initiation of premeiotic S phase. Usually this is done at 1800 hours or 2300 hours (clock time). The cultures are released on day-11 at

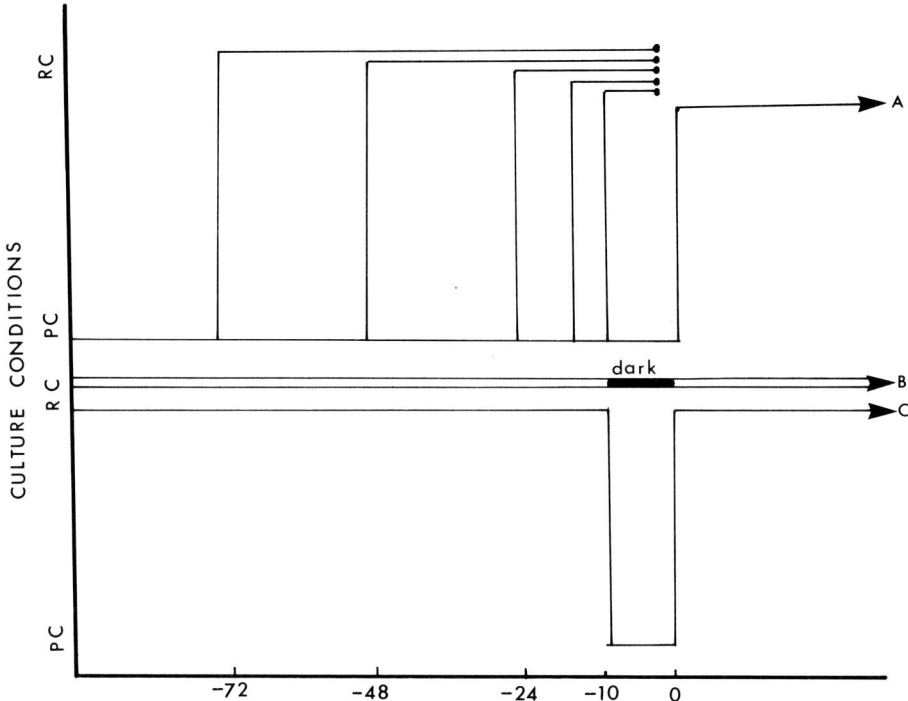

Fig. 4.2. Determination of the temperature- and light-sensitive period with respect to the initiation of meiosis as revealed by temperature shift-up, shift-down, and dark treatment experiments. A: Cultures that are shifted up to restrictive condition (RC) before the critical period cannot initiate meiosis, while those shifted up to RC after the critical period can go through meiosis normally. B: Cultures incubated in the RC can go through meiosis when the light is turned off during the critical period. C: Cultures incubated in the RC can go through meiosis when shifted down to the permissive condition (PC) during the critical period. RC = 35 °C with continuous light regime; PC = 25 °C with 16 h light–8 h dark regime.

0900 hours and, in 5–6 h, karyogamy begins irrespective of the time of release (Fig. 4.3) so long as the total time in the arrest phase does not exceed 20 h (Lu, 1974a).

The meiotic cell cycle of arrest–release cultures is most interesting. The S phase is only 5.5–6 h as compared to 8 h in the normal cultures. This may be due to the increased synchrony of the population induced by the arrest at the pre–S phase. Karyogamy and pachytene periods are unchanged, but diplotene is drastically reduced from 4.5 h to less than 1 h or may be completely absent (Lu and Chiu, 1978). We have obtained evidence that the division program is precociously induced by the effect of premeiotic S-phase arrest (see later discussions). Since the division protein(s) is ready, the cells proceed from pachytene directly into division stages (Fig. 4.4).

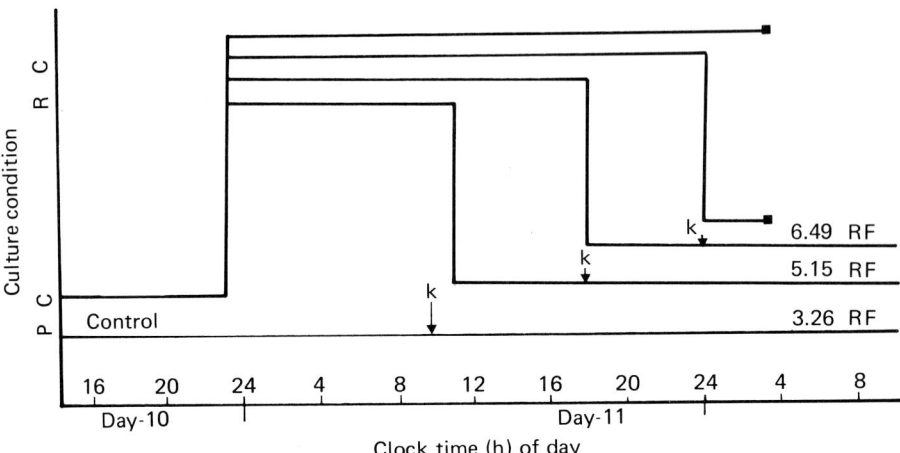

Fig. 4.3. Fruiting cultures were shifted up to restrictive condition (RC) on day-10 for the arrest of premeiotic S phase and subsequently shifted down to permissive condition (PC) at different intervals to determine the reversibility and the time of karyogamy (K-arrowed) which occurred invariably about 6 h after shift-down. The reversible period was within 20 h; prolonged arrest caused abortion of basidiocarps. The recombination frequencies (RF) from these cultures are shown on the graph.

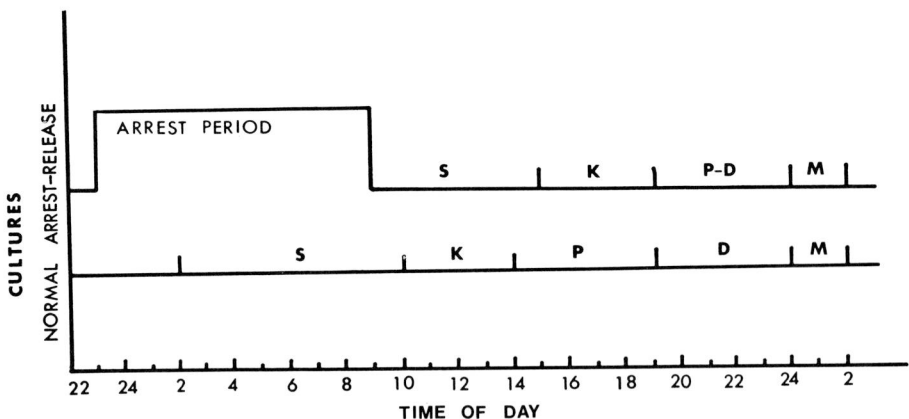

Fig. 4.4. Meiotic cycle in normal and in arrest–release cultures. The normal cultures were incubated at 25 °C under a 16 h light–8 h dark regime (PC). In the arrest-release cultures, fruiting cultures were shifted to the restrictive condition on day-10 at 2300 hours (clock time) for arrest of premeiotic S phase and were returned to the PC on day-11. The S phase began immediately upon release; karyogamy followed in 6 h. The stages of development were monitored cytologically. S, premeiotic S phase; K, karyogamy; P, pachytene; D, diplotene; P-D, pachytene–diplotene; M, meiotic divisions.

Premeiotic DNA Replication

The study of DNA replication in fungi is much more difficult than in either bacteria or higher organisms, because specific labeling of DNA with [^{3}H]-thymidine is unsuitable. However, in a synchronous system, this problem can be overcome by ^{32}P-labeling followed by chemical separation of DNA and RNA (Lu and Jeng, 1975). The incorporation kinetics of ^{32}P are shown in Fig. 4.5. It is apparent that DNA synthesis in normal cultures is complete before the onset of karyogamy. This means that the premeiotic S phase takes place before the onset of karyogamy. This situation is not unique to *Coprinus cinereus;* it has already been demonstrated by Feulgen spectrophotometry in two ascomycetes, *Neottiella rutilans* (Fr.) Dennis (Rossen and Westergaard, 1966) and *Sordaria fimicola* (Roberge) Cesati et de Notaris (Bell and Therrien, 1977).

Interestingly, in the arrest–release cultures, the incorporation kinetics are bimodel. Incorporation increases to a plateau and remains so during the arrest phase, resuming incorporation at a faster rate immediately upon release. In 6 h, DNA synthesis is complete and the fruiting bodies reach karyogamy. Inasmuch

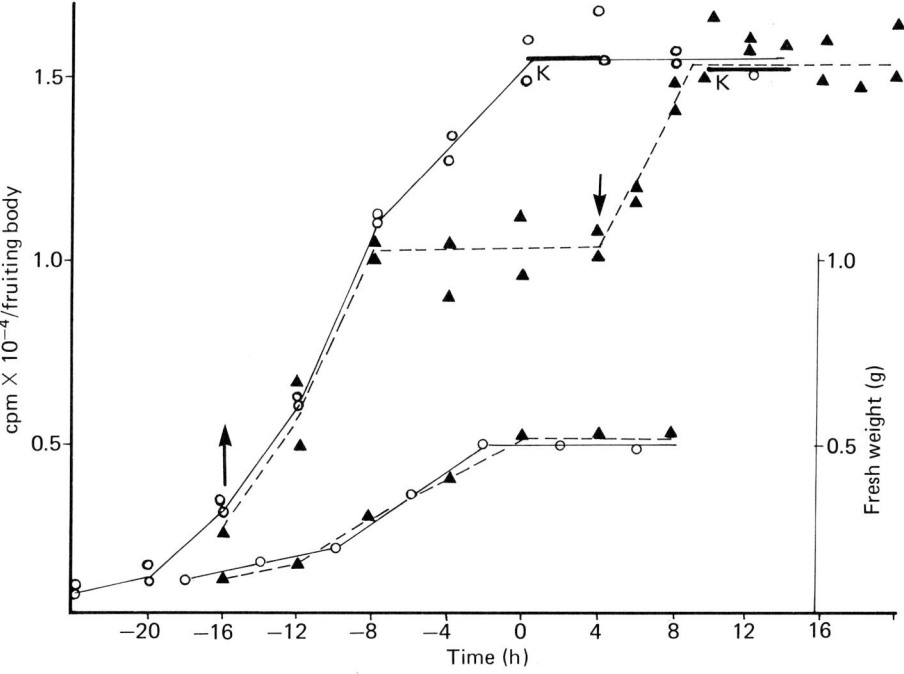

Fig. 4.5. Incorporation kinetics of ^{32}P into DNA in control and arrest–release cultures of *Coprinus cinereus:* open circles, controls; karyogamy was initiated after the incorporation had leveled off; solid triangles, arrest–release cultures. The dark bars represent the time of karyogamy (K). The arrows indicate the upward shift to the restrictive conditions and the downward shift to the permissive conditions. It is noted that only the incorporation of premeiotic DNA synthesis is arrested. The lower graph represents the fresh weight of the basidiocarps.

as confirming that the premeiotic S phase takes place before karyogamy, the incorporation kinetics of the arrest–release cultures indicate that (1) there are considerable mitotic activities before karyogamy presumably to produce more basidia before the premeiotic S phase; (2) that it takes 8 h to complete DNA replication in normal cultures and it takes only 5.5–6 h in the arrest–release cultures; and (3) that the mitotic and meiotic DNA replications are under different sets of genetic controls. The assumption that mitotic and meiotic DNA replication are under different controls is based on the fact that the mitotic DNA replication is not sensitive to high temperature and light regime as is the meiotic one.

Which stage is subject to arrest? Is it the initiation of DNA replication or is it the DNA chain elongation? Firstly, by arrest–release and re-arrest technique (Fig. 4.6), a release of 1 h or less is insufficient to allow meiosis to take place, but a release of 2 h or longer is. When this technique is combined with the ^{32}P incorporation kinetics, it becomes clear that the sensitive event involves only the initiation process of DNA replication (Fig. 4.7). Those sites (replicons) that have been initiated during the first hour of release continue to incorporate ^{32}P until completion, and those that have not been initiated during the release period are unable to initiate during the re-arrest period. As a consequence, the kinetics of a 1-h release program reaches a plateau with only half the incorporation of a 2-h release program. Also, the cultures with 1-h release fail to reach karyogamy. The DNA chain elongation is not affected by the combined effect of high temperature and

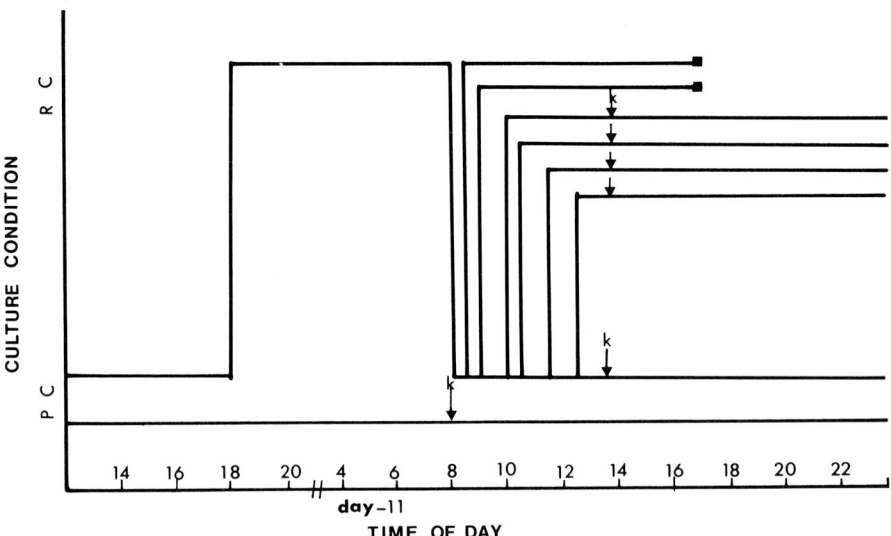

Fig. 4.6. Determination of the minimal release period required in arrest–release cultures to allow the initiation of meiosis. Cultures were shifted to the restrictive condition (RC) for the arrest of the premeiotic S phase for 14 h, and subsequently 2 cultures each were released for a varying period ranging from 0.5 h to 4.5 h to complete release. The results showed that a release of 1 h or less did not allow meiosis to take place, but 2 h or more allowed meiosis to proceed normally. k indicates the time of karyogamy.

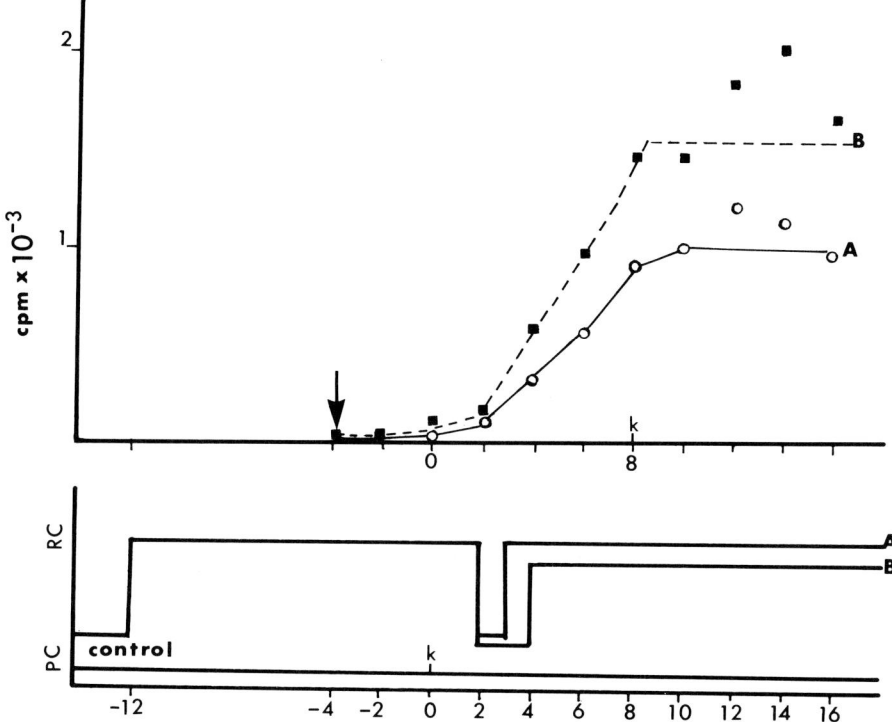

Fig. 4.7. Incorporation kinetics of ^{32}P into DNA in the arrest–release cultures of *Coprinus cinereus;* A, open circles: 1-h release; B, solid squares: 2-h release. The arrest and release scheme is shown in the lower graph. The isotope was injected into the medium 4 h before the initiation of karyogamy of the control cultures (arrowed).

light, but it can be inhibited at any time during the 8-h period of S phase by such chemicals as 2-deoxyadenosine and hydroxyurea, both of which are known inhibitors of DNA synthesis.

Genetic Recombination

Several approaches have been used to study genetic recombination in fungi. The first important contributions came from the analyses of gene conversions. The single most significant model derived from these studies is probably the hybrid DNA model of Whitehouse (1963) and Holliday (1964). The extensive research in this area has been the subject of a number of recent reviews (Emerson, 1967; Fogel and Mortimer, 1971; Stadler, 1973; Catcheside, 1974; Hastings, 1975). The second approach to this problem is the application of external stimuli to meiosing cells; the response in terms of increase or decrease in recombination frequency can be used as an indicator of the recombination mechanisms involved. In this presentation, I will focus on this latter approach.

Effect of High Temperature on Recombination

The use of temperature as a probe was first adopted by geneticists studying higher organisms such as *Drosophila* (Plough, 1917; Grell, 1966, 1973), grasshoppers (Henderson, 1970; Peacock, 1970), and maize (Maguire, 1968). The same was applied with equal success in fungi (McNelly-Ingle et al., 1966; Towe and Stadler, 1964; Landner, 1970; Stamberg and Simchen, 1970; Lu, 1969, 1970, 1974b; Lu and Chiu, 1976; Raju and Lu, 1973). The results from *Coprinus cinereus* are shown in Fig. 4.8.

The recombination response to high temperature is not stage-specific; it can take effect at late S phase, at karyogamy, and at pachytene. On the other hand, the response to low temperature is very stage-specific; only at pachytene can a maximum increase of recombination be effected by cold treatment. This suggests that high and low temperature probably affect different steps in the recombination process (Raju and Lu, 1973). The following experiments seem to confirm that this is the case.

Studies in prokaryotes indicate that DNA repair processes are intimately involved in genetic recombination (Bernstein, 1968; Krisch et al., 1972; Zieg and

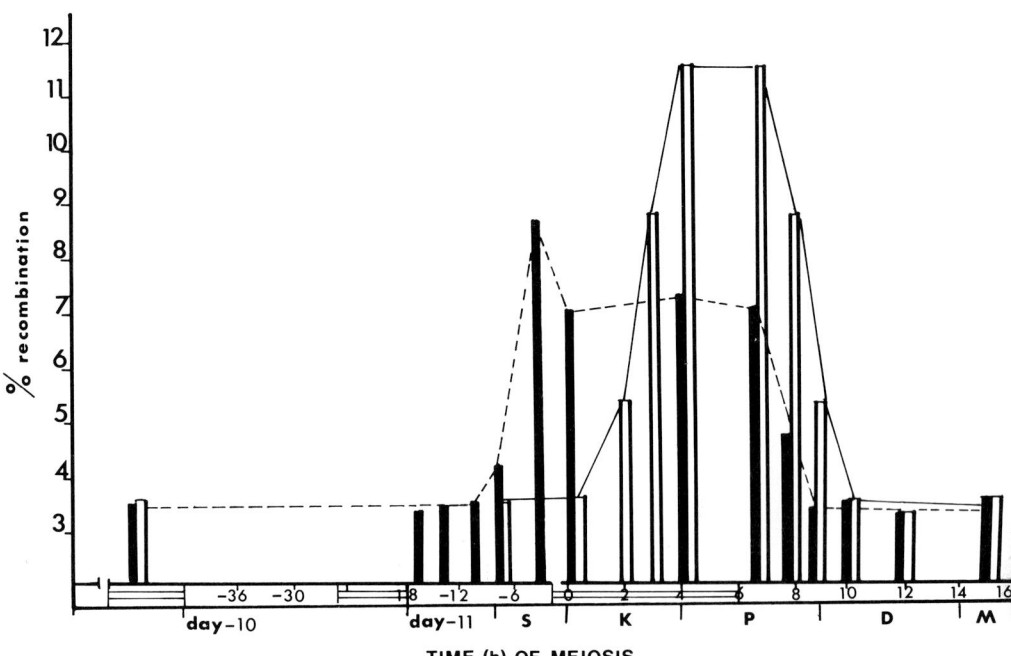

Fig. 4.8. A summary of the temperature-sensitive periods during which recombination frequency between *den* and *met-1* can be increased by high and low temperatures. The sensitive periods at karyogamy and pachytene were reported earlier (Lu, 1969). Solid histogram, 3 h heat treatment at 35 °C; open histogram, 13 h cold treatment at 5 °C. Note the different time scales. Abbreviations as in Fig. 4.4.

Kushner, 1977; and the reviews by Clark, 1973, Radding, 1973, Broker, 1975, and Eisenstark, 1977). I believe that there are two opposing processes: nicking and repairing. If nicks are repaired immediately, the chance for these nicks to be involved in crossing over is very small. On the other hand, if nicks are left open, the chances are increased that they will become involved in crossing over. The balance between nicking and repairing would then determine the frequencies of crossing over. If this hypothesis is correct, then recombination should be increased by increasing nicks or decreasing repair activities.

Since *Coprinus cinereus* has a synchronous meiotic system, it is possible to make a preliminary investigation into the repair activities during meiotic prophase by using ^{32}P-labeling experiments. (Due to technical problems common in fungal cells, more definitive studies of nicking and repairing cannot be achieved at present.) The results are shown in Fig. 4.9. Clearly, the peak repair activity in terms

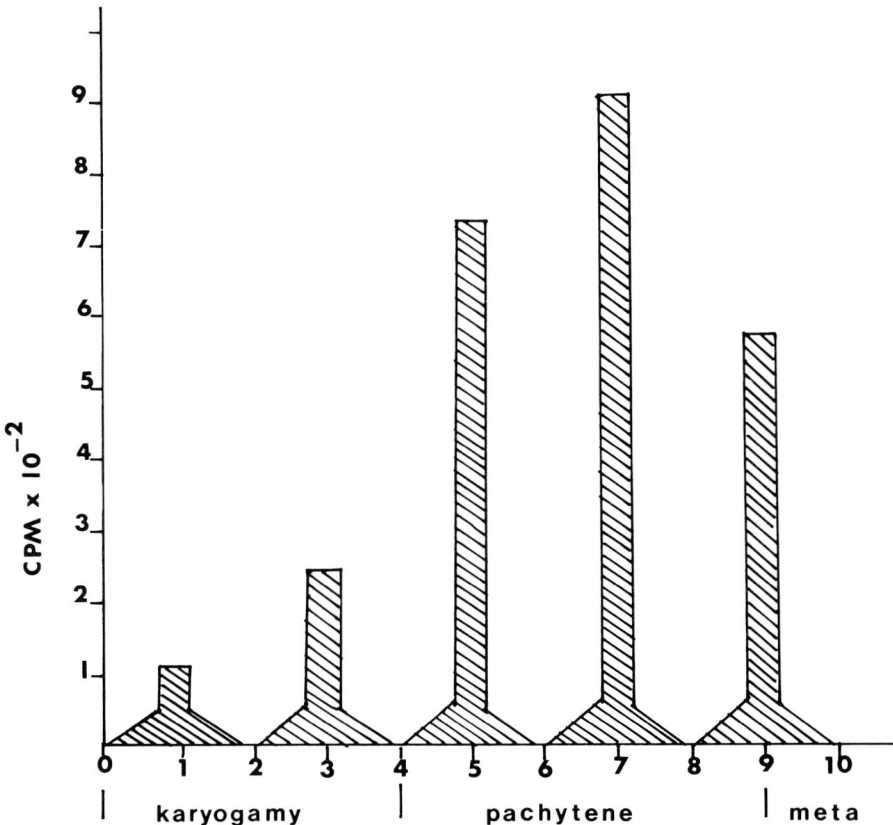

Fig. 4.9. The rate of repair DNA synthesis during meiosis as shown by ^{32}P pulse-labeling experiment. Cultures were synchronized by the arrest–release technique. At the onset of karyogamy (0 time) 50 μCi of [^{32}P] phosphoric acid was injected into the medium of each culture at 2-h intervals for a 2-h pulse-labeling until metaphase I was reached. DNA was extracted from 4 fruiting bodies and counted in a liquid scintillation counter. The results show that the repair activity peaks at late pachytene.

of ^{32}P incorporation is found at late pachytene in the arrest–release cultures. When the ^{32}P incorporation experiment is combined with high temperature treatment, the results are quite revealing. As shown in Fig. 4.10, high temperature treatment at pachytene does not increase incorporation as compared with the controls (A and B in Fig. 4.10). This suggests that the rate of repair activity is about the same with or without the heat treatment. However, when pulse labeled at the period immediately after heat treatment (to be called recovery period), there is a 3-fold increase in the treated cultures over the controls (C and D in Fig. 4.10). This finding suggests that high temperature produces more nicks or gaps in the DNA which call for additional repair activities.

If one can assume that the increased ^{32}P incorporation is a reflection of increased nicking, the explanation for the recombination response to high temperature becomes obvious. Any nicks or gaps induced before the onset of repair activity will remain open and hence will increase the chance of crossing over. Thus, heat treatment at late S phase, at karyogamy, or at pachytene, will lead to an increased recombination frequency. This conclusion is based on the hypothesis

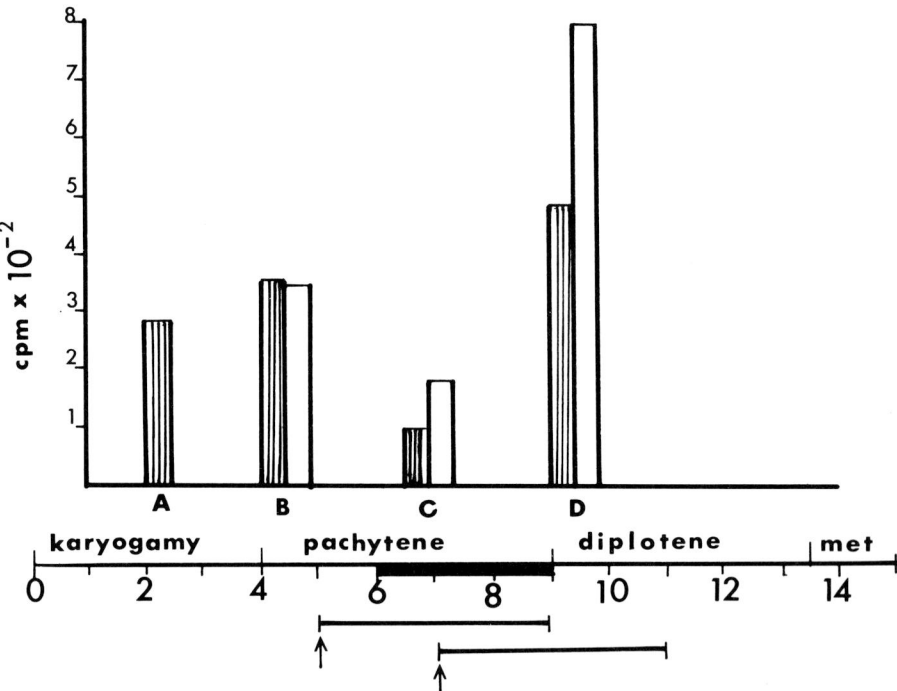

Fig. 4.10. Repair synthesis as indicated by incorporation of ^{32}P into DNA from heat treatment experiments in normal cultures. The periods of labeling and the stages involved are shown at the lower part of the graph. The 3-h heat treatment was done between 6 and 9 h of postkaryogamy time. The arrows indicate the time of the injection of ^{32}P, which takes 2 h to reach the meiotic cells. A: 2 h pulse-labeling at pachytene; B: same as in A with 3 h heat treatment at 35 °C; C: 2 h pulse at early diplotene; D: same as in C with 3 h heat treatment at 35 °C.

Replication of DNA and Crossing Over in *Coprinus* 103

that meiotic repair activity is not turned on until pachytene. The evidence provided in Fig. 4.9 appears to support this. More definitive data in this regard are being sought at present.

Effect of Cold Temperature on Recombination

The cold temperature effect on recombination is very stage-specific (Fig. 4.8). It is effective only if treatment is given at pachytene (Lu, 1969) when synaptonemal complexes are present (Lu, 1970). Further studies have shown that recombination can also be increased in arrest–release cultures. If cold treatments are applied at hourly intervals beginning 4 h after the onset of karyogamy, it is apparent that a maximum increase can be induced by such treatment even at late pachytene (Fig. 4.11). How does the cold treatment bring about an increase in recombination? I have already shown that the repair activity is peaked at late pachytene (Fig. 4.9), and this peak period happens to coincide with the latest period when a maximum increase of recombination can be induced (Fig. 4.11). This may be more than just a coincidence, and could be interpreted to mean that high repair activity will reduce the number of gaps in the DNA which are potential spots in which crossing over can take place and consequently decrease the chance of recombination.

Thus, to understand the effect of cold treatment on recombination, it seems

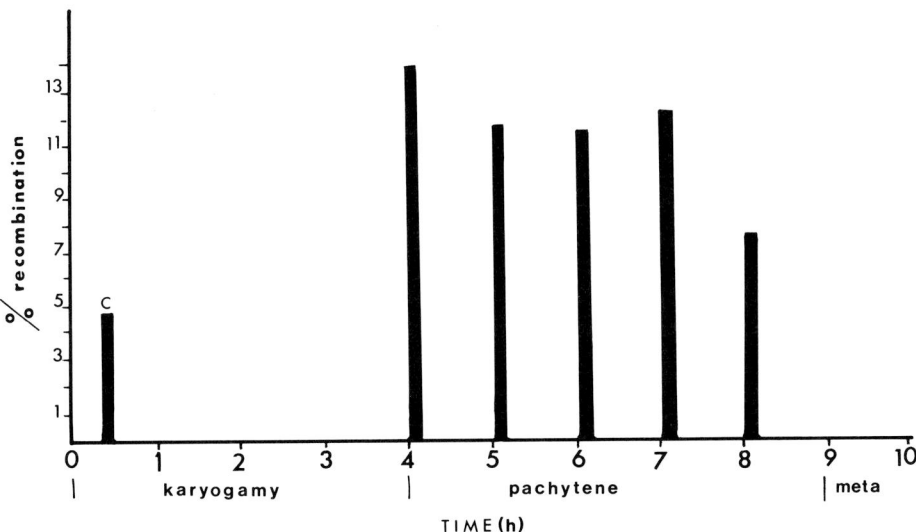

Fig. 4.11. Effect of cold temperature on recombination frequency (RF) in arrest–release cultures of *Coprinus cinereus*. The RF is increased by 13 h of treatment at 5 °C during early and late pachytene (between 4 and 8 h postkaryogamy time). The onset of karyogamy is designated as 0 time and the period of karyogamy was determined by light microscopy. The duration of pachytene is also confirmed by the presence of synaptonemal complexes. It is clear that diplotene is drastically reduced from 4.5 h in normal cultures to less than 1 h in arrest–release cultures. C: control.

logical to examine the repair activity during the cold treatment period. As shown in Fig. 4.12, the repair activity is drastically suppressed. It is also found that, given a recovery period, the total ^{32}P incorporation is the same as the controls (Fig. 4.12). Apparently, cold treatment does not increase the number of gaps in the DNA; therefore, no increase in ^{32}P incorporation occurs. Cold treatment simply suppresses the repair activity, and the existing gaps need to be repaired during recovery period. The recombination event has been fully committed by the end of the treatment (see Fig. 4.14 for evidence); the recovery repair activity, though necessary, is not involved in the commitment to crossing over. This finding is in perfect agreement with my earlier hypothesis that recombination can be increased by decreasing repairs. The hypothesis can be extended to suggest that when nicks or gaps in DNA are not repaired there is a better chance for crossing over to take place. The suppression of repair activity may promote crossing over only when the homologous chromosomes are locked together by the synaptonemal complex. For this reason, one might expect the cold temperature effect to be pachytene-specific, which is, in fact, the case. The results also indicate that the actual crossover event is fully committed by late pachytene.

The effect of cold treatment on meiosis has been examined, in arrest–release cultures, and the results shed some light on the overall picture of the recombination process. As shown in Fig. 4.13, fruiting bodies given a 13-h cold treatment at the beginning of pachytene (i.e., 4 h after the onset of karyogamy) require 5 h to reach metaphase I, whereas those given the same at the end of pachytene (i.e., 8

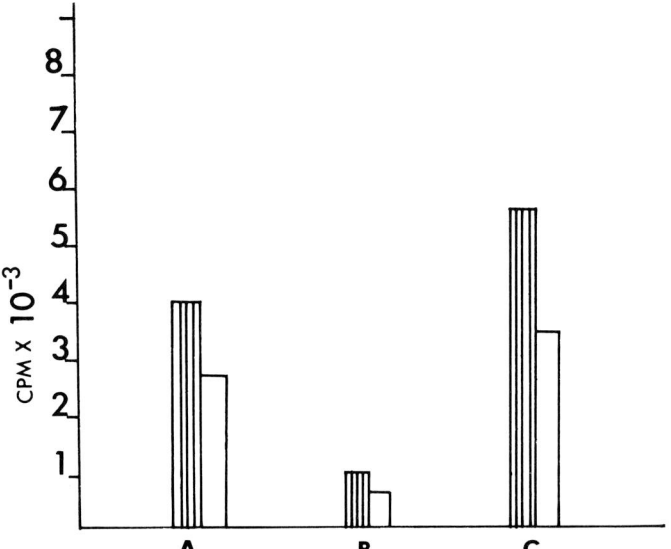

Fig. 4.12. Repair synthesis as indicated by incorporation of ^{32}P into DNA from cold-treatment experiment in arrest–release cultures of *Coprinus cinereus*. The shaded and the open histograms represent 2 replicates. A: control 3 h labeling at pachytene; B: 7 h labeling during the cold treatment period which resulted in 3-fold increase in recombination; C: 11 h labeling during 7 h cold treatment plus 4 h recovery period.

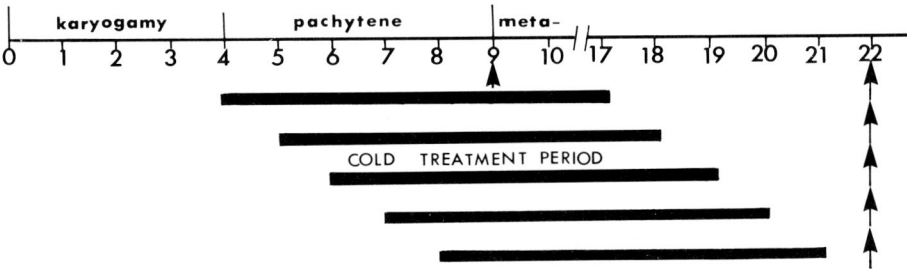

Fig. 4.13. Effect of cold treatment on the progress of meiosis in arrest–release cultures of *Coprinus cinereus*. Fruiting cultures were given 13 h of cold treatment in hourly intervals during pachytene. The recovery time to reach metaphase I in each culture was monitored. Irrespective of the beginning time of the cold treatment, all cultures reached metaphase I 13 h after the control cultures had reached metaphase I. This shows that the cultures stand still during the cold-treatment period.

h after the onset of karyogamy) require only 1 h. All cultures, irrespective of the time at which cold treatments were given, reach metaphase I at the same time (i.e., 13 h after the controls have reached metaphase I). The results indicate that the cold temperature has simply suspended the development of meiosis, but allows the crossing over commitment to take place (see evidence below in Kinetic Study of Temperature Effect on Recombination). This commitment is believed to involve Holliday's (1964) heteroduplex formation. If this assumption is accepted, the formation of Holliday's structure can take place at 5°C with little or no DNA synthesis.

Kinetic Study of Temperature Effect on Recombination

The results obtained above have shown that recombination appears to be facilitated by holding two chromosomes together for a longer time and by reducing repair activities. These results suggest that either the recombination process or its commitment is time dependent. This suggestion can be tested by kinetic studies.

Fruiting bodies at pachytene were given from 1 to 16 h of cold temperature treatment; the results are shown in Fig. 4.14. It is clear that the kinetics do not fit a linear (i.e., single-hit) function; rather, they fit a quadratic (i.e., two-hit) function, and the optimal time is 7 h (Lu, 1974b). Such a curve suggests: (1) that the recombination frequency is a function of time: the longer the two homologous chromosomes are held together by the synaptonemal complex, the higher the recombination frequency; (2) that at the end of cold treatment all recombination events are committed irreversibly. The recovery repair activity discussed earlier does not influence those recombination sites already committed; rather, it prevents those uncommitted sites from becoming committed.

From the above evidence and discussions, one can formulate the following hypothesis concerning the mode in which recombination occurs. There are a finite number of sites that may be involved in crossing over. The distribution of these

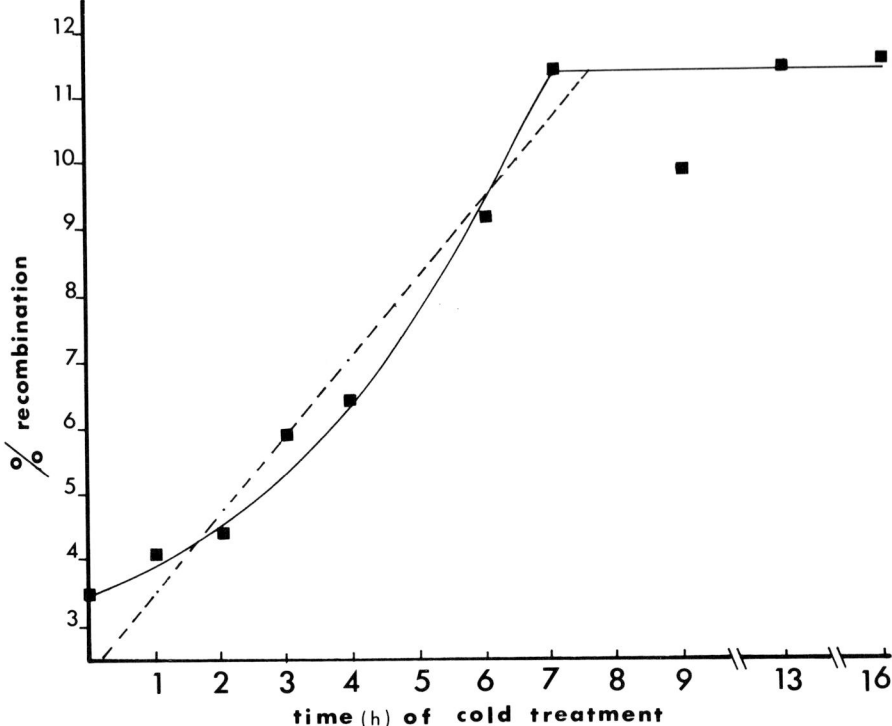

Fig. 4.14. Fruiting bodies at the beginning of pachytene were given from 1 to 16 h of cold treatment at 5 °C, and the recombination frequencies (RF) were scored. Each solid square represents an average of 2–5 independent treatments. The data cannot be adequately explained by a linear regression ($P < 0.01$), which is shown in the dotted line. The 9 h of cold treatment gave a subnormal RF because only 80% of basidia completed karyogamy at the time of treatment.

sites along the chromosomes are not necessarily random. Some regions may have more sites than others. However, within a given region, such as between *den* and *met-1* of chromosome III of *Coprinus cinereus,* nicks at these sites along two homologous chromosomes are random. When identical sites on both chromosomes are nicked (to be called coincidental nicks) a crossover potential is created. Given time, such potentials will lead to crossing over, and hence recombination.

Under normal cultural conditions, nicks caused at these sites by endonuclease activities are repaired, and the chance of coincidental nicks is very much reduced. Since the endonuclease activities and repair activities in a species are constant, the number of coincidental nicks at any given region are also constant; therefore, a constant map distance can be estimated. If fruiting bodies are given a heat treatment which causes more nicks on the chromosomes, the chance for coincidental nicks is increased; consequently, the recombination frequency is also increased (Lu, 1969, 1974b). If the nicking activity is constant as in normal cultures, but the fruiting bodies are given a cold treatment at pachytene before the major repair activity becomes functional, all coincidental nicks are maintained. All crossover potentials then become fully committed to crossing over [most probably by Holliday's (1964) heteroduplex formation] if given sufficient time, and the maximum recombination frequency is achieved. At this point, longer treatment time will lead to no further increase in the recombination frequency.

Since coincidental nicks are required, a two-hit kinetics is expected. This hypothesis is difficult to prove experimentally but a computer simulation supports the argument.

The computer was programmed to make random nicks on the homologous chromosomes within a region including 100 possible sites for crossing over. Whenever there is a coincidental nick, crossing over may occur. The computer was also programmed to assume that 4 nicks on each chromosome are involved in the first hour of cold treatment, 8 nicks in the second hour, 12 nicks in the third hour, etc., for a total of 7 h. The computer prediction and the actual cold-treatment data are shown in Fig. 4.15. From this exercise, it is clear that the experimental data is in perfect agreement with the computer predictions. Thus the results support the hypothesis proposed above.

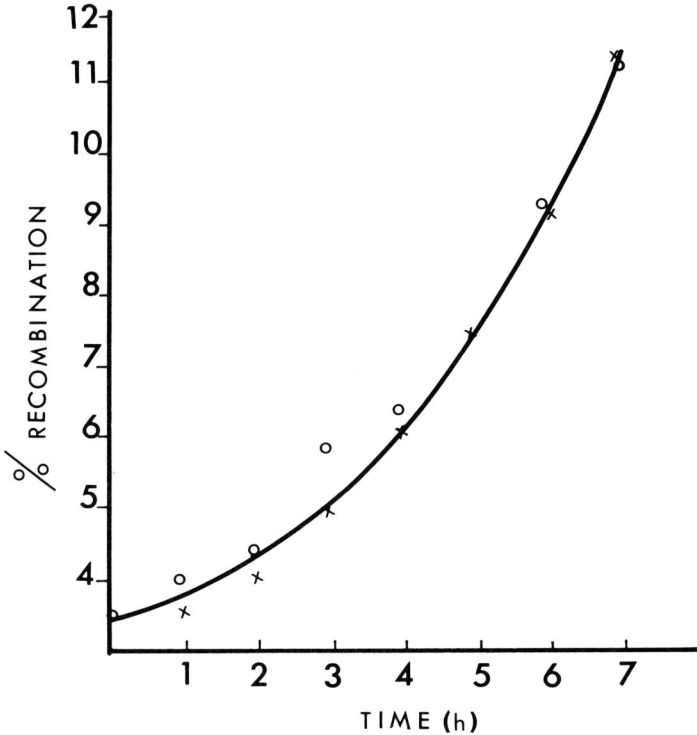

Fig. 4.15. A computer-simulated kinetic of cold treatment effect on recombination based on the hypothesis proposed in the text together with the experimental data (Lu, 1974b). The two are in perfect agreement. The solid line is the computer best fit for a quadratic function; the crosses are expected frequency of recombination based on the hypothesis; the open circles are actual data of recombination induced by different length of time of cold treatment.

Recombination, a Coordinated Program of the Meiotic Cell Cycle

Recombination through crossing over is too important an event to be left to chance alone. It therefore seems logical to presume that evolution has taken a course which ensures that such events take place in every meiosing cell by programming recombination into the meiotic cell cycle. The following thoughts are quite speculative, although there are enough pieces of the jigsaw that appear to fit together to outline a believable story. I will first outline what regulatory programs appear to be concerned and then try to fill in the evidence I have so far.

As shown in Fig. 4.16, I propose that there is a nicking program which prepares single-stranded gaps as a necessary precondition for crossing over to take effect. This program begins at late S phase and ends at late pachytene. The temporal limits of this period are defined by the period of increased recombination by high temperature (Lu, 1969, 1974b). My reasons for proposing the nicking program are as follows: (1) Evidence discussed earlier in this presentation shows that high temperature increases gaps on DNA (Lu and Chiu, 1976). (2) I propose that high temperature enhances the activity of the nicking enzyme (this is as yet unproven). Whenever this enzyme is present the nicking program is present. High-tempera-

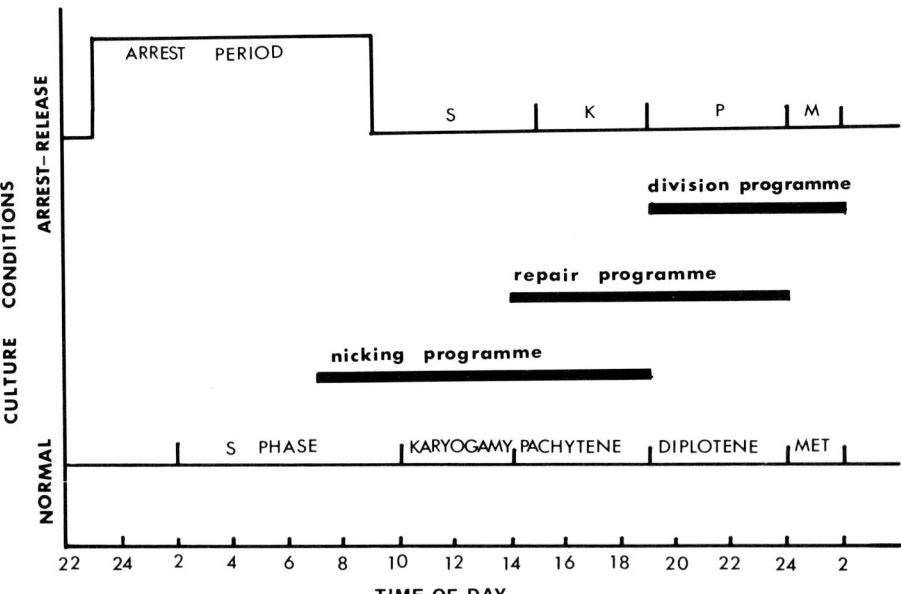

Fig. 4.16. Cellular programs of the meiotic cell cycle in normal and in arrest–release cultures of *Coprinus cinereus*. The temporal appearance of nicking, repair, and division programs are indicated by the dark bars. The programs are coordinated with the meiotic cell cycle in normal cultures. However, in the arrest–release cultures, they are believed to take place unaltered even though the DNA-division cycle has been arrested. As a result, they are dissociated from the nuclear division cycle (see text for detailed explanation). For example, the division program takes place at pachytene instead of at diplotene. S, premeiotic S phase; K, karyogamy; P, pachytene; M or MET, meiotic divisions.

ture treatment causes an increase in recombination frequency as long as the treatment is given before the occurrence of the actual crossover events.

I also propose that there is a repair program that does not come into effect until mid- to late pachytene. This is based on the labeling experiment discussed earlier in this presentation. Delay in turning on the repair program has the distinct advantage of being able to maximize recombination, because the single-stranded gaps created by the nicking program may be left open and then have the chance to cross over when the chromosomes are held together by the synaptonemal complexes.

What evidence do I have that there are such programs and that they are coordinated during the meiotic cell cycle? In a recent paper (Lu and Chiu, 1978), I have suggested that there is a nuclear cycle and a cytoplasmic cycle in which cellular programs are made. The two cycles are independent but are normally coordinated during meiosis. When the nuclear cycle is arrested, the cytoplasmic programs continue; as a result, the two cycles become disjointed. The first evidence that this suggestion is correct is provided by the experiments involving the division program (Lu and Chiu, 1978). When the nuclear cycle is arrested, the cytoplasmic programs continue as scheduled. As a result, the division program becomes precociously induced in the arrest–release cultures. Since the division program is ready, diplotene is dispensed with or drastically reduced from 4.5 h to less than 1 h. As shown in Fig. 4.16, the division program is completed on schedule even though the nuclear cycle is arrested, and the completed program falls at the time of pachytene of the arrest–release cultures.

If the recombination is part of a coordinated program governing the meiotic cell cycle, disruption of the cell cycle should also influence recombination events. To examine the nicking program, I have again used high temperature as a probe. As shown previously (Lu, 1974b), heat treatment does not cause an increase in recombination until late S phase and continues to be effective until late pachytene in normal cultures. In the arrest–release cultures, however, the arrest by high temperature (Fig. 4.3) causes an increase of recombination before S phase. In other words, the nicking program becomes precociously induced; the longer the arrest, the higher the recombination frequency (Lu, 1974b). This can only be explained by the fact that when the nuclear cycle is arrested the nicking program has already been scheduled. As shown in Fig. 4.16, the nicking program falls betwen pre–S phase and the end of karyogamy in the arrest–release cultures. If the nicking program is shifted to an earlier part of the nuclear cycle, it should also terminate precociously. I examined this possibility by a series of heat treatments at hourly intervals throughout the meiotic stages (Lu, unpublished) and found this to be so. As demonstrated in Fig. 4.17., the period of increased recombination induced by heat treatment stops at the beginning of pachytene in the arrest–release cultures as contrasted to the results of the normal cultures.

In conclusion, the cellular events involved in meiotic recombination in *Coprinus cinereus* appear to be part of a coordinated program governing the meiotic cell cycle. Among the obvious, there is a nicking program and a repair program. The balance of the two brings about the frequency of recombination. The two programs are staggered in time so as to maximize the number of recombination events.

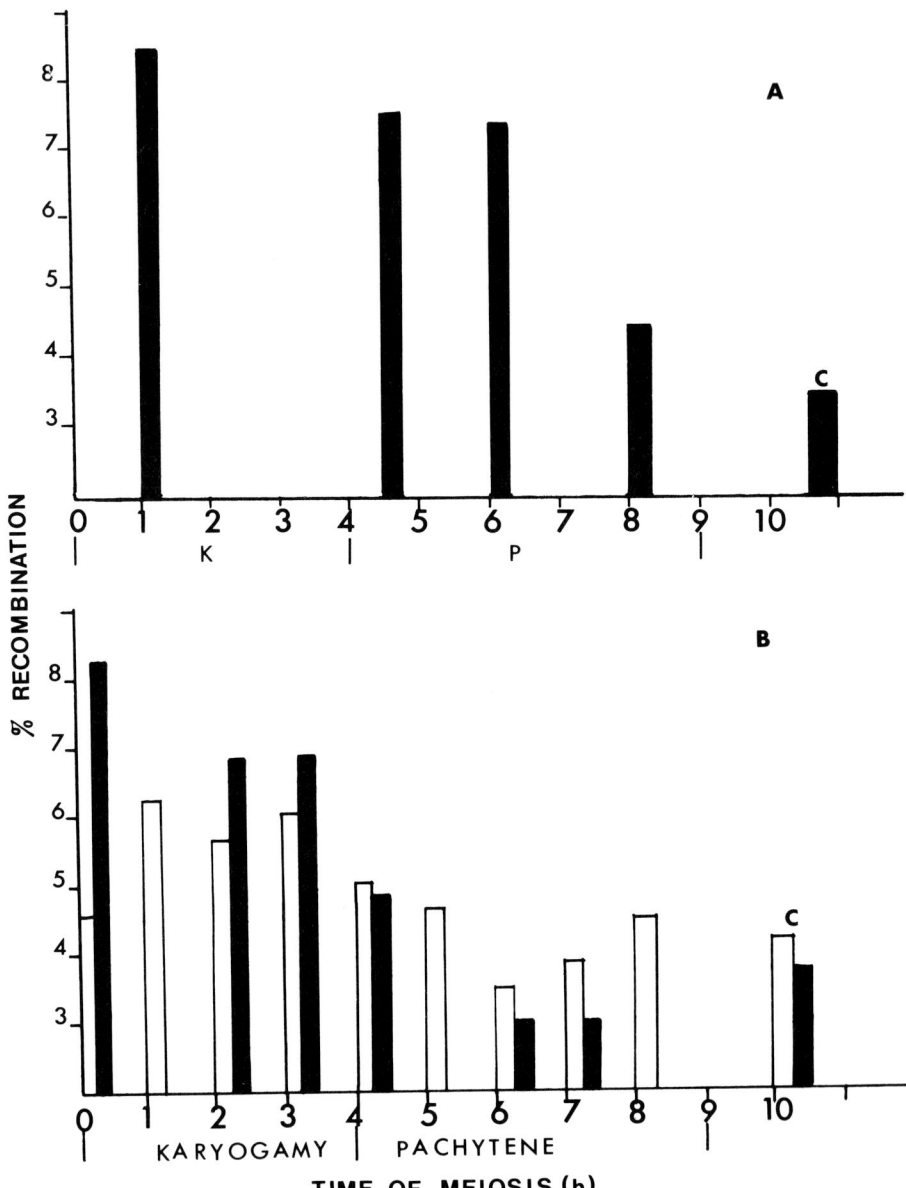

Fig. 4.17. Effect of 3 h of treatment at 35 °C on genetic recombination in the cross *den* × *met-1* of *Coprinus cinereus*. Recombination can be increased at early and late pachytene in normal cultures (A) but not in arrest–release cultures (B).

Concluding Remarks

Coprinus cinereus possesses one of the few synchronous meiotic systems that has been studied in some detail. The advantages of such a synchronized system have been amply demonstrated in studies of meiosis and genetic recombination. It should be recognized that it is not a system without disadvantages. But when the difficulties, common to fungi, have been overcome, a great deal more can be learned about meiosis and recombination in this system.

The use of external stimuli to probe genetic recombination has been shown to be as valuable as other approaches in bringing about understanding of genetic recombination. A similar approach has also been used quite fruitfully to shed light on the mechanism of recombination in prokaryotes (Eisenstark, 1977). All lines of evidence appear to point to the intimate involvement of nicking and repairing programs of the meiotic cell cycle in the events of genetic recombination.

Acknowledgments. I wish to thank the National Research Council of Canada, Ottawa, for financial assistance.

References

Bell, W. R., Therrien, C. D.: A cytophotometric investigation of the relationship of DNA and RNA synthesis to ascus development in *Sordaria fimicola*. Can. J. Genet. Cytol. *19*, 359–370 (1977).

Bernstein, H.: Repair and recombination in phage T4. I. Genes affecting recombination. Cold Springs Harb. Symp. Q. B. *33*, 325–331 (1968).

Broker, T. R.: Molecular and genetic recombination of bacteriophage T4. Ann. Rev. Genet. *9*, 213–244 (1975).

Catcheside, D. G.: Fungal Genetics. Ann. Rev. Genet. *8*, 279–300 (1974).

Clark, A. J.: Recombination deficient mutants of *E. coli* and other bacteria. Ann. Rev. Genet. *7*, 67–86 (1973).

Eisenstark, A.: Genetic recombination in bacteria. Ann. Rev. Genet. *11*, 369–396 (1977).

Emerson, S.: Fungal genetics. Ann. Rev. Genet. *1*, 201–220 (1967).

Fogel, S., Mortimer, R. K.: Recombination in yeast. Ann. Rev. Genet. *5:* 219–236 (1971).

Grell, R. F.: The meiotic origin of temperature-induced crossovers in *Drosophila melanogaster* females. Genetics *54*, 411–421 (1966).

Grell, R. F.: Recombination and DNA replication in the *Drosophila melanogaster* oocyte. Genetics *73*, 87–108 (1973).

Hastings, P. J.: Some aspects of recombination in eukaryotic organisms. Ann. Rev. Genet. *9*, 129–144 (1975).

Henderson, S. A.: The time and place of meiotic crossing-over. Ann. Rev. Genet. *4*, 295–324 (1970).

Holliday, R.: A mechanism for gene conversion in fungi. Genet. Res. Camb. *5*, 282–304 (1964).

Krisch, H. M., Hamlett, N. V., Berger, H.: Polynucleotide ligase in bacteriophage T4D recombination. Genetics *72*, 187–203 (1972).

Landner, L.: Variation of recombination frequency in *Neurospora crassa* following temperature changes prior to and during meiosis and evidence for a premeiotic sensitive stage. Molec. Gen. Genet. *109*, 219–232 (1970).

Lu, B. C.: Genetic recombination in *Coprinus:* I. Its precise timing as revealed by temperature-treatment experiments. Can. J. Genet. Cytol. *11*, 834–847 (1969).

Lu, B. C.: Genetic recombination in *Coprinus:* II. Its relations to the synaptonemal complexes. J. Cell Sci. *6*, 669–687 (1970).

Lu, B. C.: Dark dependence of meiosis at the elevated temperatures in the basidiomycete *Coprinus lagopus*. J. Bacteriol. *111*, 833–834 (1972).

Lu, B. C.: Meiosis in *Coprinus:* VI. The control of the initiation of meiosis in *C. lagopus*. Can. J. Genet. Cytol. *16*, 155–164 (1974a).

Lu, B. C.: Genetic recombination in *Coprinus:* IV. A kinetic study of the temperature effect on recombination frequency. Genetics *78*, 661–677 (1974b).

Lu, B. C., Chiu, S. M.: Genetic recominbation in *Coprinus:* V. Repair synthesis of deoxyribonucleic acid and its relation to meiotic recombination. Mol. Gen. Genetics *147*, 121–127 (1976).

Lu, B. C., Chiu, S. M.: Meiosis in *Coprinus:* IX. The influence of premeiotic S phase arrest and cold temperature on the meiotic cell cycle. J. Cell Sci. *32*, 21–30 (1978).

Lu, B. C., Jeng, D. Y.: Meiosis in *Coprinus:* VII. The prekaryogamy S phase and the postkaryogamy DNA replication in *C. lagopus*. J. Cell Sci. *17*, 461–470 (1975).

Maguire, M. P.: Evidence on the stage of heat induced crossover effect in maize. Genetics *60*, 353–362 (1968).

McNelly-Ingle, C. A., Lamb, B. C., Frost, L. C.: The effect of temperature on recombination frequency in *Neurospora crassa*. Genet. Res. *7*, 169–183 (1966).

Peacock, W. J.: Replication, recombination, and chiasmata in *Goniaea australostae* (Orthoptera: Acridiae). Genetics *65*, 593–617 (1970).

Plough, H. H.: The effect of temperature on crossing over. J. Exp. Zool. *32*, 187–212 (1917).

Radding, C. M.: Molecular mechanisms in genetic recombination. Ann. Rev. Genet. *7*, 87–111 (1973).

Raju, N. B., Lu, B. C.: Meiosis in *Coprinus:* III. Timing of meiotic events in *C. lagopus* (sensu Buller). Can. J. Bot. *48*, 2183–2186 (1971).

Raju, N. B., Lu, B. C.: Genetic recombination in *Coprinus:* III. Influence of gamma-irradiation and temperature treatment on meiotic recombination. Mutation Res. *17*, 37–48 (1973).

Rossen, J. M., Westergaard, M.: Studies on the mechanism of crossing over. II. Meiosis and the time of meiotic chromosome replication in the Ascomycete *Neottiella rutilans* (Fr.) Dennis. C. R. Trav. Lab. Carlsberg *35*, 233–260 (1966).

Stadler, D. R.: The mechanisms of intragenic recombination. Ann. Rev. Genet. *7*, 113–128 (1973).

Stamberg, J., Simchen, G.: Specific effects of temperature on recombination in *Schizophyllum commune*. Heredity *25*, 41–52 (1970).

Towe, A. M., Stadler, D. R.: Effects of temperature on crossing over in *Neurospora*. Genetics *49*, 577–583 (1964).

Whitehouse, H. L. K.: A theory of crossing-over by means of hybrid DNA. Nature, Lond. *199*, 1034–1040 (1963).

Zieg, J., Kushner, S.: Analysis of genetic recombination between two partially deleted lactose operons of *Escherichia coli* K-12. J. Bacteriol. *131*, 123–132 (1977).

Chapter 5

Biochemical and Genetic Studies on the Initial Events of Fruitbody Formation

ISAO UNO AND TATSUO ISHIKAWA

Introduction

The process of fruitbody formation in higher basidiomycetes involves the expression of a series of structural and regulatory genes required for the morphogenetic reactions. The first reaction necessary to initiate fruiting may be triggered by a genetic factor under certain environmental conditions. The incompatibility factors have been studied extensively since it has been shown that these factors regulate dikaryotization prerequisite to fruiting in the normal process of development (Raper, 1966; Raper and Raper, 1968). It is, however, known that fruiting is not limited to dikaryotic mycelia; monokaryotic mycelia produce fruitbodies under particular conditions such as aging, injury, or influence of some substances (Stahl and Esser, 1976). Various types of mutants that form fruitbodies on monokaryotic mycelia have also been found (Stahl and Esser, 1976). Study of monokaryotic fruiting has the advantage of dealing with a simple system of initiating fruiting free from the combination of incompatibility factors. We have made a series of experiments showing that adenosine 3′,5′-cyclic monophosphate (cyclic AMP) is one of the trigger substances in monokaryotic fruiting of *Coprinus macrorhizus* Rea f. *microsporus* Hongo. The present report describes the outline of the work and suggests a possible mechanism through which cyclic AMP exerts the effect on fruiting.

Detection of Fruitbody-Inducing Substance

We have found that a particular monokaryotic strain (fis^+ strain) of *C. macrorhizus* produced monokaryotic fruitbodies by the addition of crude extract of dikaryotic fruitbodies of *C. macrorhizus* or other species such as *Lentinus edodes* (Berk.) Sing. and *Tricholoma mutsutake* (Ito et Imai) Sing. (Fig. 5.1 and Table 5.1). From the extensive studies to identify the fruitbody-inducing substances involved in the crude extracts of fruitbodies, we concluded that the substances that are effective in inducing monokaryotic fruitbodies in the fis^+ strain are cyclic AMP and adenosine 3′-monophosphate (3′-AMP) (Uno and Ishikawa, 1973a,

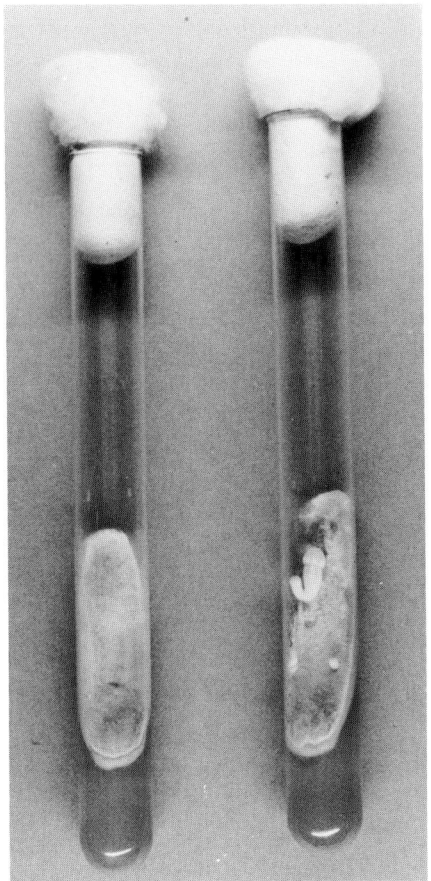

Fig. 5.1. Monokaryotic cultures of fis^+ strain of *Coprinus macrorhizus* grown for 10 days with (right) or without (left) crude extract prepared from fruitbodies of *Lentinus edodes*.

1973b). Cyclic AMP and 3′-AMP added in the culture media were quite active in inducing fruiting in the fis^+ strain (Table 5.1). However, mycelia of other monokaryotic strains (fis^- strains) formed no fruitbody even with sufficient supply of crude extracts of fruitbodies or cyclic AMP (Table 5.1).

The fis^+ strain was not stable but frequently produced a fruitbody that resulted in a culture (fis^c strain) producing monokaryotic fruitbodies constitutively without addition of crude extract of fruitbodies or cyclic AMP (Table 5.1). The experimental results indicated that a significant amount of fruitbody-inducing substances was included in mycelia of the fis^c strain and the dikaryon that are able to form fruitbodies, but not in those of fis^- strains that are unable to form fruitbodies (Uno and Ishikawa, 1971). These facts suggest that the fruitbody-inducing substances found may participate as a common chemical basis for the induction of fruiting in this fungus.

Table 5.1. Induction of Monokaryotic Fruitbodies in Various Strains of *Coprinus macrorhizus* (Uno and Ishikawa, 1971, 1973b).

		Addition[a]						
		Crude extract of fruitbodies				3'-AMP		
Strain	None	*Coprinus Macrorhizus*	*Lentinus edodes*	*Tricholoma matsutake*	Cyclic AMP (20 nM)	(20 nM)	Caffeine (20 nM)	Theophylline (20 nM)
Monokaryon								
fis⁻	−	−	−	−	−	−	−	−
fis⁺	±	+	+	+	+	+	+	+
*fis*ᶜ	+	+	+	+	+	+	+	+
Dikaryon[b]	+	+	+	+	+	+	+	+

[a] At least 10 independent slant cultures of a strain were made and incubated at 30°C for 15 days. +, fruitbodies were observed in most cultures tested; −, no fruitbody was observed in all cultures tested; ±, fruitbodies were observed in one or two slant cultures tested.
[b] Dikaryon prepared between strains *fis*⁻ A8B7 and *fis*⁻ A7B8.

Regulation of Cyclic AMP Level and Fruiting

The *fis*ᶜ strain of *C. macrorhizus* was able to synthesize [³H]-cyclic AMP from [³H]-adenine in the culture medium, but the *fis*⁻ strain failed to synthesize it under the same condition (Uno and Ishikawa, 1973a). It has been known that cyclic AMP is synthesized from adenosine triphosphate (ATP) by adenylate cyclase and degraded to adenosine 5'-monophosphate (5'-AMP) by phosphodiesterase. The activities of these enzymes were detected in monokaryotic mycelia of *fis*ᶜ and *fis*⁺ strains and in dikaryotic mycelia that were able to form fruitbodies, but not in those of *fis*⁻ strains that are unable to form fruitbodies (Table 5.2).

Mycelia of *fis*⁺ strain contained approximately 10 times as much activity of phosphodiesterase as mycelia of *fis*ᶜ strain and the dikaryon (Table 5.2). It has been speculated that the *fis*⁺ mycelia containing a large amount of phosphodiesterase rapidly degrade cyclic AMP synthesized by adenylate cyclase; and, therefore, the *fis*⁺ mycelia require an external supply of cyclic AMP to form fruitbodies. Further study indicated that phosphodiesterase activity is inhibited by 3'-AMP as well as by caffeine and theophylline that are known as potent inhibitors of cyclic nucleotide-specific phosphodiesterase, but not by 5'-AMP or other nucleotides (Table 5.3). These inhibitors of phosphodiesterase were active in inducing monokaryotic fruitbodies by the addition to the *fis*⁺ cultures (Table 5.1). The results indicate that 3'-AMP, caffeine, and theophylline were active in inducing fruitbodies in the *fis*⁺ strain by inhibiting extraordinarily high phosphodiesterase activity, which digested cyclic AMP synthesized by adenylate cyclase in the mycelium.

Mycelia of *fis*⁻ strains produce no or very low activities of adenylate cyclase and phosphodiesterase. The external supply of cyclic AMP to the *fis*⁻ cultures was not effective in inducing fruitbodies in the *fis*⁻ strains (Table 5.1). This may be explained by the fact that mycelia of *fis*⁻ strains were unable to incorporate a

Table 5.2. Levels of Adenylate Cyclase, Phosphodiesterase, and Cyclic AMP in the 8-Day-Old Mycelia of 6 Strains of *Coprinus macrorhizus* Grown under Various Conditions (Uno and Ishikawa, 1973b, 1974b; Uno et al., 1974).

Strain	Glucose concn (%)	Growth condition	Fruitbody formation Normal	Fruitbody formation Malformed[a]	Adenylate cyclase[b]	Phosphodiesterase[b]	Cyclic AMP level[c]
Dikaryon[d]	0.4	light	+	−	210	0.25	7.7
	0.4	dark	−	+	110	0.17	−
	5.0	light	−	−	50	0.12	1.4
fis^-	0.4	light	−	−	0	0	0.8
	0.4	dark	−	−	0	0	−
	5.0	light	−	−	0	0	−
fis^+	0.4	light	±	−	105	1.36	−
fis^c	0.4	light	+	−	120	0.20	7.8
	0.4	dark	−	−	0	0	0.6
	5.0	light	−	−	30	0.11	1.9
ds	0.4	light	+	−	68	0.15	−
	0.4	dark	−	+	88	0.10	−
gluR	0.4	light	+	−	115	0.12	6.7
	5.0	light	+	−	110	0.17	6.5

[a]Malformed fruitbodies have long stripes and small undeveloped pilei.
[b]The specific activities (units/mg protein) of adenylate cyclase and phosphodiesterase in crude mycelial extracts were measured by the methods of Khandelwal and Hamilton (1971) and Butcher and Sutherland (1962), respectively.
[c]The cyclic AMP levels (nmol/g dry mycelium) were measured by the immunological assay of Steiner et al. (1972) or the protein binding assay of Gilman (1970). −, not tested.
[d]Dikaryon prepared between strains fis^- A8B7 and fis^- A7B8.

significant amount of [^{3}H]-cyclic AMP added to the culture medium, although mycelia of fis^+ and fis^c strains were able to incorporate it (Table 5.4).

Regulation of fruitbody formation and cyclic AMP level in mycelial cells by environmental conditions has also been studied. As shown in Table 5.2, mycelia of fis^c strain grown in the dark formed no fruitbodies and produced extremely low

Table 5.3. Effect of Various Compounds on Phosphodiesterase Activity of *Coprinus macrorhizus* (Uno and Ishikawa, 1973b).

Addition	Concn (μM)	Phosphodiesterase activity[a] (% of control)
None	—	100
Caffeine	2	0
Theophylline	2	0
3′-AMP	10	82
	20	52
5′-AMP	20	100
3′-GMP	20	96
5′-GMP	20	96

[a]Phosphodiesterase activity was assayed by a method in which [^{3}H]-cyclic AMP was used to differentiate the reaction product from nucleotide added in the reaction mixture (Uno and Ishikawa, 1973b).

Table 5.4. Summary of Fractionation of Crude Extracts Prepared from Mycelia of 3 Strains of *Coprinus macrorhizus* After Incorporation of [^{3}H]-Cyclic AMP (Uno and Ishikawa, 1973b).

Fraction[a]	^{3}H-radioactivity (10^4 cpm)		
	fis^+	fis^c	fis^-
Incorporated [^{3}H]-cyclic AMP (total)	75	85	1.7
Trichloroacetic acid-soluble fraction	57.4	47.8	1.5
ATP + ADP	35.1	32.9	0
cyclic AMP	6.2	4.7	1.4
AMP	0.1	0.1	0
Trichloroacetic acid-insoluble fraction	18.5	12.5	0
RNA	9.3	7.5	0
DNA	8.4	5.5	0

[a] The 8-day-old cultures were incubated with [^{3}H]-cyclic AMP (20 μCi) for 10 h. Mycelia thus treated were fractionated as described by Uno and Ishikawa (1973b).

level of cyclic AMP (Uno et al., 1974). A short illumination of fis^c mycelia grown in the dark resulted in the formation of fruitbody primordia and syntheses of adenylate cyclase and phosphodiesterase (Uno et al., 1974). A mutant strain (*ds* strain) that formed malformed fruitbodies in the dark was able to synthesize significant amounts of adenylate cyclase and phosphodiesterase even in the dark (Table 5.2).

Fruitbody formation and accumulation of cyclic AMP in mycelia of the fis^c strain and in the dikaryon were equally inhibited in the media containing more than 3% glucose (high glucose medium; Table 5.2). This suggests that the catabolite repression is observed in the fruitbody formation of this fungus (Uno and Ishikawa, 1974b). A mutant strain (*gluR* strain) which formed fruitbodies on the high glucose medium produced higher levels of cyclic AMP as compared with the fis^c strain or the dikaryon grown in the same medium (Table 5.2).

The observations indicate that fruitbodies are formed only under the conditions in which cyclic AMP is accumulated in mycelial cells.

Existence of Cyclic AMP-Dependent Protein Kinases

The occurrence of protein kinase that is activated by cyclic AMP has been found by Walsh et al. (1968) in rabbit skeletal muscle, and similar enzymes have since been reported to be present in many eukaryotic cells (Rubin and Rosen, 1975). It is now believed that cyclic AMP-activated protein kinases are receptors for cyclic AMP in various tissues, and are responsible for biological effects of this nucleotide (Rubin and Rosen, 1975). In *C. macrorhizus,* we have identified at least three species of protein kinase on a Sepharose 6B chromatogram of mycelial

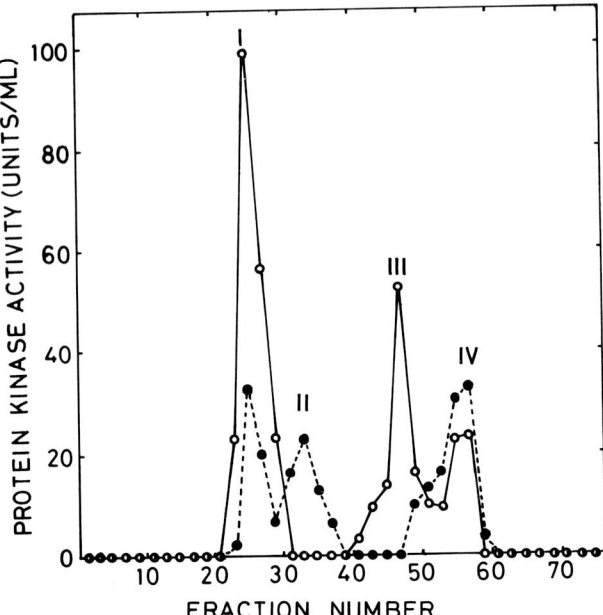

Fig. 5.2. Sepharose 6B chromatogram of protein kinases in mycelial extract of fis^c strain of *Coprinus macrorhizus* grown in the light for 8 days (Uno and Ishikawa, 1974a). Each fraction was assayed for the protein kinase activity with histone as substrate in the presence (●) and absence (○) of 1 μM cyclic AMP. I–IV indicate major peaks of the activity.

extract of fis^c strain (Fig. 5.2). One species of the enzyme (peak II) was active only in the presence of cyclic AMP. In contrast, other species of the enzyme (peak III) was inhibited by the addition of cyclic AMP. The third species of the enzyme (peak IV) showed no significant response to cyclic AMP, although there is a possibility that this type of enzyme is the catalytic unit of cyclic AMP-dependent protein kinase. Histone, casein, and serum albumin were equally well phosphorylated with these enzymes (Uno and Ishikawa, 1974a).

It has been shown that the level of protein kinase activity was dependent on the culture age of mycelia and elution patterns of protein kinases on Sepharose 6B chromatogram were different between fis^c mycelia grown in the light and those grown in the dark (Uno and Ishikawa, 1975). It may be, therefore, predicted that the difference in the species and amount of protein kinases may reflect on the ability of fruitbody formation.

Regulation of Glycogen Metabolism by Cyclic AMP

It has been well documented in mammalian tissues that cyclic AMP-dependent protein kinases are responsible for the interconversion of active and inactive forms of glycogen synthetase and glycogen phosphorylase (Rubin and Rosen, 1975; Billar-Palasi and Larner, 1970). In *C. macrorhizus*, the addition of cyclic AMP to partially purified mycelial extracts of fis^c strain resulted in the inhibition of gly-

Table 5.5. Effect of Cyclic AMP, ATP, and Glucose 6-Phosphate on Glycogen Synthetase and Glycogen Phosphorylase Activities in *Coprinus macrorhizus* (Uno and Ishikawa, 1976, 1978).

Addition			Enzyme activity (% of control)[b]	
Cyclic AMP (5.0 μM)	ATP (3.3 mM)	G-6-P[a] (2.5 mM)	Glycogen synthetase	Glycogen phosphorylase
−	−	−	100	100
+	−	−	9	150
−	+	−	101	109
+	+	−	2	300
−	−	+	98	−
+	−	+	49	−
+	+	+	51	−

[a] G-6-P, glucose 6-phosphate.
[b] Crude extract prepared from *fis*[c] mycelia grown in the light was filtered through a Sephadex G-25 column (2 × 40 cm), and the turbid fractions were used as the enzyme sample. Enzyme activities were assayed in the presence (+) or absence (−) of additives as indicated by the methods described by Uno and Ishikawa (1976, 1978).

cogen synthetase and activation of glycogen phosphorylase (Table 5.5, Fig. 5.3). Glucose 6-phosphate was effective in reversing the inhibition of glycogen synthetase by cyclic AMP. ATP was effective in stimulating the activation of glycogen phosphorylase by cyclic AMP. The levels of cyclic AMP effective in inhibiting or activating the enzymes were approximately equal to the level of cyclic AMP in the mycelia (Table 5.2, Fig. 5.3).

A cellular fraction responsible for the activation of glycogen phosphorylase has been found in a Sepharose 6B chromatogram of mycelial extract of *fis*[c] strain (Fig. 5.4). The fraction contained cyclic AMP-dependent protein kinase activity and cyclic AMP-binding activity (Fig. 5.4). Such a fraction responsible for the inactivation of glycogen synthetase has not been found in Sepharose 6B chromatogram of the same kind of mycelial extract. The fact that glycogen phosphorylase was activated with cyclic AMP in the presence of ATP and a cellular fraction containing the protein kinase suggests that the phosphorylation of the enzyme may be responsible for the activation as indicated in mammalian systems.

The activation of glycogen phosphorylase with cyclic AMP and ATP was observed in the cellular extract of *fis*[c] mycelia grown in the light that are able to form fruitbodies but not in those of *fis*[c] mycelia grown in the dark and *fis*[−] mycelia grown in the light that are unable to form fruitbodies (Uno and Ishikawa, 1976). These results indicate that a proper protein kinase to activate glycogen phosphorylase may not be involved in mycelia which are unable to form fruitbodies.

In mycelia of the strain *fis*[c] grown in the light which are able to form fruitbodies, the glycogen synthetase activity was high at the early phase of growth and began to decrease 5 days after inoculation, but the glycogen phosphorylase activity was at a maximum 6 days after inoculation and decreased thereafter (Uno and Ishikawa, 1976, 1978). On the other hand, the accumulation of cyclic AMP

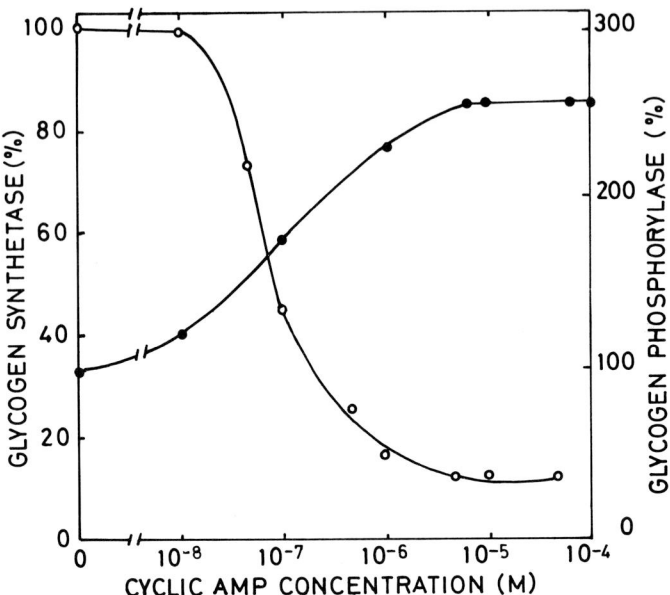

Fig. 5.3. Effect of cyclic AMP on the activities of glycogen synthetase and glycogen phosphorylase (Uno and Ishikawa, 1976, 1978). Crude extract prepared from fis^c mycelia of *Coprinus macrorhizus* grown in the light was filtered through a Sephadex G-25 column and the turbid fractions were used as the enzyme sample (Sephadex G-25 fraction). Enzyme activities were assayed with the indicated amount of cyclic AMP and percent of the activity to the control (without cyclic AMP) was described. ○, Glycogen synthetase activity; ●, glycogen phosphorylase activity assayed with ATP (3.3 mM).

in fis^c mycelia reached a maximum level 6 days after inoculation, and the amount of glycogen in mycelial cells of this strain was at the maximum level 6 days after inoculation and rapidly decreased thereafter (Uno and Ishikawa, 1976). These data may be consistent with a scheme that the cyclic AMP synthesized regulates the amount of glycogen in mycelia-producing fruitbodies by the inhibition of glycogen synthetase and the activation of glycogen phosphorylase.

Possible Role of Cyclic AMP in Fruiting

On the basis of our experiments described above, we propose a hypothesis that cyclic AMP acts as a trigger substance on fruitbody formation in *Coprinus macrorhizus*. The incompatibility factors, regulatory genes such as fis^+, fis^c, and *ds*, and the environmental factors such as light and high glucose concentration may regulate the level of cyclic AMP by controlling the activities of adenylate cyclase and phosphodiesterase simultaneously. Heteroallelic combinations of incompatibility factors in dikaryotic cells permit the synthesis of cyclic AMP. The fis^c mutation will be effective to derepress the synthesis of adenylate cyclase and

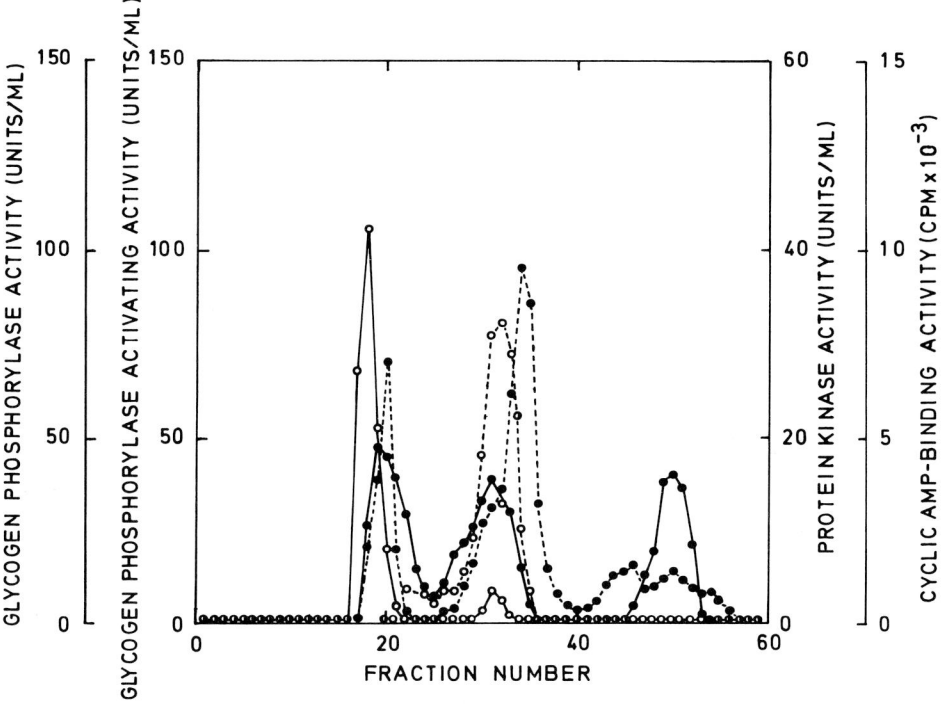

Fig. 5.4. Sepharose 6B chromatogram of glycogen phosphorylase, glycogen phosphorylase-activating activity, cyclic AMP-dependent protein kinase, and cyclic AMP-binding activity in the Sephadex G-25 fraction prepared from fis^c mycelia of *Coprinus macrorhizus* grown in the light (see Fig. 5.3). Experimental methods were described by Uno and Ishikawa (1976). o———o, Glycogen phosphorylase activity; o---o, glycogen phosphorylase activating activity; •———•, protein kinase activity assayed with 1 μM cyclic AMP; •---•, cyclic AMP-binding activity.

phosphodiesterase in the light, and the light-requiring process may be replaced by the *ds* mutation.

Catabolite repression has been observed in this species, and the synthesis of inducible enzymes such as D-serine deaminase and tryptophanase was repressed with higher concentration of glucose (Uno and Ishikawa, 1974b). The positive regulation of protein synthesis as observed in the operon systems of bacteria may operate in this fungus, and a possibility exists that cyclic AMP may exhibit the activity by binding with a part of operons and stimulating the protein synthesis necessary for fruiting. However, a more attractive scheme for the action of cyclic AMP in *C. macrorhizus* may be that through cyclic AMP-dependent protein kinases, as well known in mammalian systems. Cyclic AMP accumulated in mycelial cells may bind with a particular protein and activate protein kinases that phosphorylate proteins responsible for the reactions leading to the fruitbody formation. At the same time cyclic AMP may inhibit other kinds of protein kinases that are required to operate only during the vegetative growth.

The candidate of protein species which are phosphorylated by the cyclic AMP-dependent protein kinase may be proteins directly involved in the fruiting process as shown for the enzymes of glycogen metabolism described in this report. It would, however, be appropriate to point out that the candidates of such protein species may well be those related to important biological activities such as ribosomal proteins, histones, and proteins constituting microtubules and cellular membranes.

References

Billar-Palasi, C., Larner, J.: Glycogen metabolism and glycolytic enzymes. Annu. Rev. Biochem. *39*, 639–672 (1970).

Butcher, R. W., Sutherland, E. W.: Adenosine 3',5'-phosphate in biological materials. I. Purification and properties of cyclic 3',5'-nucleotide phosphodiesterase and use of this enzyme to characterize adenosine 3',5'-phosphate in human urine. J. Biol. Chem. *237*, 1244–1250 (1962).

Gilman, A. G.: A protein binding assay for adenosine 3',5'-cyclic monophosphate. Proc. Natl. Acad. Sci. USA *67*, 305–312 (1970).

Khandelwal, R. L., Hamilton, I. R.: Purification and properties of adenyl cyclase from *Streptococcus salivarius*. J. Biol. Chem. *246*, 3297–3304 (1971).

Raper, J. R.: Genetics of Sexuality in Higher Fungi. New York: Ronald Press 1966.

Raper, J. R., Raper, C. A.: Genetic regulation of sexual morphogenesis in *Schizophyllum commune*. J. Elisha Mitchell Sci. Soc. *84*, 267–273 (1968).

Rubin, C. S., Rosen, O. M.: Protein phosphorylation. Annu. Rev. Biochem. *44*, 831–887 (1975).

Stahl, U., Esser, K.: Genetics of fruitbody production in higher basidiomycetes. I. Monokaryotic fruiting and its correlation with dikaryotic fruiting in *Polyporus ciliatus*. Molec. Gen. Genet. *148*, 183–197 (1976).

Steiner, A. L., Parker, C. W., Kipnis, D. M.: Radioimmunoassay for cyclic nucleotides. I. Preparation of antibodies and iodinated cyclic nucleotides. J. Biol. Chem. *247*, 1106–1113 (1972).

Uno, I., Ishikawa, T.: Chemical and genetical control of induction of monokaryotic fruiting bodies in *Coprinus macrorhizus*. Molec. Gen. Genet. *113*, 228–239 (1971).

Uno, I., Ishikawa, T.: Purification and identification of the fruiting inducing substances in *Coprinus macrorhizus*. J. Bacteriol. *113*, 1240–1248 (1973a).

Uno, I., Ishikawa, T.: Metabolism of adenosine 3',5'-cyclic monophosphate and induction of fruiting bodies in *Coprinus macrorhizus*. J. Bacteriol. *113*, 1249–1255 (1973b).

Uno, I., Ishikawa, T.: Presence of multiple protein kinase activities in *Coprinus macrorhizus*. Biochim. Biophys. Acta *334*, 354–360 (1974a).

Uno, I., Ishikawa, T.: Effect of glucose on the fruiting body formation and adenosine 3',5'-cyclic monophosphate level in *Coprinus macrorhizus*. J. Bacteriol. *120*, 96–100 (1974b).

Uno, I., Ishikawa, T.: A biochemical basis of fruiting body formation in *Coprinus macrorhizus*. Proc. 1st. Inter. Congr. IAMS (Tokyo) *1*, 313–321 (1975).

Uno, I., Ishikawa, T.: Effect of cyclic AMP on glycogen phosphorylase in *Coprinus macrorhizus*. Biochim. Biophys. Acta *452*, 112–120 (1976).

Uno, I., Ishikawa, T.: Effect of cyclic AMP on glycogen synthetase in *Coprinus macrorhizus*. J. Gen. Appl. Microbiol. *24,* 193–197 (1978).
Uno, I., Yamaguchi, M., Ishikawa, T.: The effect of light on fruiting body formation and adenosine 3′,5′-cyclic monophosphate metabolism in *Coprinus macrorhizus*. Proc. Natl. Acad. Sci. USA *71,* 479–483 (1974).
Walsh, D. A., Perkins, J. P., Krebs, E. G.: An adenosine 3′,5′-monophosphate dependent protein kinase from rabbit skeletal muscle. J. Biol. Chem. *243,* 3763–3765 (1968).

Chapter 6

Control of Stipe Elongation by the Pileus and Mycelium in Fruitbodies of *Flammulina velutipes* and Other Agaricales

HANS E. GRUEN

Introduction

Interactions between different regions of a mycelium and between different mycelia play an important role in morphogenesis and growth of fungi. The better understood interactions are those which are mediated by substances secreted into the substrate, and which result in the initiation and growth of sexual organs. In *Achlya* and Mucorales specific sexual hormones have been identified (Barksdale, 1969; Bu'Lock, 1976; van den Ende, 1976), and diffusible, but as yet unidentified compounds play a similar role in the sexual interactions of *Ascobolus stercorarius* (Bull.) Schroet. (Bistis, 1956, 1957). Bistis also provided strong evidence that initiation of apothecia depends on a diffusible agent released by the ascogonium.

Compared to the evidence for the morphogenetic role of secreted substances much less is known about interactions between different regions of individual hyphae or of hyphal associations. Butler (1961) noted interdependence in elongation between branches on a main hypha, and Larpent (1966) showed that initiation and elongation of branch hyphae is influenced by apical dominance in the parent hypha. Nutrient levels in the substrate, translocation of cell contents towards the main apex, and internal competition between the main hypha and its branches participate in these interactions but the underlying mechanism remains to be clarified. The small size and mode of growth of individual hyphae present considerable obstacles to the study of internal growth correlations. Such studies can be carried out more readily on large hyphal associations represented by fruitbodies of many higher fungi, expecially those of hymenomycetes. Differentiation and enlargement of these fruitbodies occur in an orderly manner, and must result from highly coordinated growth of numerous interconnected hyphae. This type of development virtually demands the existence of internal coordinating mechanisms. Physiological relationships among different fruitbody regions with tissue-like organization can be investigated macroscopically. Such studies have been done mainly on fruitbodies of fleshy, lamellate Agaricales which have finite growth. Nothing is known about the physiological interactions in primordia during differentiation of the fruitbody regions when cell enlargement is limited. The present review is concerned with the subsequent transitional period which is marked by an increasing rate of cell enlargement, and with the phase of rapid fruitbody enlargement when cell elongation in the stipe attains its maximum rate. The role of the pileus in stipe elongation will be emphasized, and participation of the mycelium in this relationship will be considered. The review of published literature will be supplemented by results of recent, largely unpublished work on *Flammulina velutipes* (Curt. ex Fr.) Sing.

Role of the Pileus in Stipe Elongation of Agaricales

As early as 1842, Schmitz mentioned that complete decapitation of young agaric fruitbodies always resulted in cessation of stipe growth, but that they grew to their "destined height" even if half of their pilei were removed. He was also the first to determine by measurement that the growth zone is located in the upper portion of the stipe. Gräntz (1898) reported that stipes of *Coprinus sterquilinus* (Fr.) Fr. grew normally after being decapitated shortly before the onset of rapid elongation, and that detached whole pilei or segments expanded normally. Also the excised growth zone of the stipe was said to elongate in the usual manner. Gräntz thought that a nutritional relationship between pileus and stipe during differentiation caused subsequent coordinated expansion of these regions. He further concluded that connection of the stipe with the mycelium was no longer required during elongation provided that sufficient water was available. Continued growth and geotropic curvature of detached stipes with or without pilei, of longitudinal stipe slices, and of stipe sections were also observed by Knoll (1909) in a *Coprinus* species (probably misidentified) during rapid elongation. Streeter's (1909) observations on transplanted stipes of *Amanita crenulata* Peck also seemed to indicate that removal of the pileus did not prevent continued elongation and geotropic response. On the other hand, Magnus (1906) mentioned that elongation of young fruitbodies of *Agaricus campestris* [probably *A. bisporus* (Lange) Imbach][1] continued if at least half the pileus was left on the stipe, but only occasionally after decapitation during rapid elongation. None of the authors cited measured the growth after various operations or presented quantitative comparisons with untreated controls.

The first quantitative results were published by Borriss (1934a) for *C. lagopus* [probably *C. radiatus* (Bolt. ex Fr.) S. F. Gray (Pinto-Lopes and Almeida, 1972)]. He showed that elongation soon ceased if stipes were decapitated before the onset of rapid elongation (stage III), but it continued if only a thin pileus slice with few lamellae was left on the stipe at an even earlier stage (II). Small curvatures away from the slice occurred often. Growth also continued if the stipe was decapitated just before the beginning of rapid elongation (stage IV). Measurements were given for only one specimen for each operation. Borriss stated that

---

[1] More than one *Agaricus* species has been cultivated and the nomenclature is very confusing. It is often impossible to determine with certainty which species was used, especially in older experimental studies.

elongation of stipes with few remaining lamellae may have been less than normal, but that he did not have enough data for adequate comparison with the variable final length of intact fruitbodies. Also, he did not mention whether stipes which were decapitated during rapid elongation attained the same final length as intact fruitbodies. He was the first to suggest possible hormonal control of stipe growth by the pileus. However, he did not obtain continued growth of young decapitated stipes by replacing mature but closed pilei on the apex, or by injecting press juice from such pilei into the stipe cavity. He thought that these negative results indicated that the age of the two structures could be a critical factor. Jeffreys and Greulach (1956) reported that rapidly growing *C. sterquilinus* stipes continued to elongate and responded geotropically after decapitation, but did not curve when agar blocks with an unspecified stipe extract were placed asymmetrically on the cut end, or when a solution was sprayed on one side. Removal of half the pileus gave no curvature during rapid elongation.

Urayama (1956) removed the lamellae from one side of wedge-shaped pileus remnants in young *Agaricus bisporus* fruitbodies and observed that they curved away from the side with the remaining lamellae. Small portions of lamellae applied on agar blocks to one side of the pileus remnants freed of lamellae also induced curvature away from that side. Similar results were reported by Hagimoto and Konishi (1959) who also found that stipes in the early stage of rapid growth elongated only little after decapitation while older ones "grew considerably." They confirmed Urayama's finding that the lamellae influenced pileus expansion. Removal of half the pileus from young fruitbodies, but not later during rapid elongation, caused curvature in three *Coprinus* species and in *Psathyrella disseminata* Pers. Later (1960) they obtained stipe curvatures indicative of growth promotion in response to lamellar diffusate applied unilaterally to pileus remnants and directly to the stipe. Diffusate from the pileus context and from stipes were also said to cause negative curvatures. Active material was extractable from lamellae with water, ethanol, ether, and acetone, and resisted boiling for 1 h in water and to some extent in N HCl and N NaOH. It diffused through cellophane and was light stable. Ethanol extracts from pilei of *C. macrorhizus* Rea f. *microsporus* Hongo, *Hypholoma fasciculare* [= *Nematoloma fasciculare* (Hudson ex Fr.) Karst.], and from lamellae of *Armillaria matsutake* [= *Tricholoma matsutake* (S. Ito et Imai) Sing.] were said to be active in the *Agaricus* curvature test. No quantitative data were given for the above findings. Hagimoto's (1963) measurements showed a decrease in stipe curvature away from lamellae left on one side of pileus remnants as the operation was performed on older fruitbodies. Also, older stipes curved up to 30° both towards (positive) and away (negative) from the side where lamellar diffusate was applied to pileus remnants. With concentrated extract, curvatures were mostly positive or changed from positive to negative, while diluted extract gave mostly negative curvatures. No measurements were given for the tests with extracts, and the significance of the curvatures cannot be evaluated. During preparation of test stipes these workers sliced surface portions from opposite sides of the stipes along their entire length. Such injury disturbs the differential tissue tensions which exist across agaric stipes (Knoll, 1909; Buller, 1909, p. 42) and can induce erratic curvatures or modify expected curvatures (see discussion in Gruen, 1967).

Hagimoto and Konishi interpreted all their results in terms of a hypothetical hormone. Hagimoto (1963) explained the positive stipe curvatures by postulating that lamellar remnants, diffusates, and extracts supplied superoptimal amounts of hormone to the stipes when they reached a certain stage of rapid elongation. Hagimoto (1964) also remarked that excess growth hormone may be partly responsible for the finite growth of fruitbodies. Konishi and Hagimoto (1962) and Konishi (1967) then reported that ethanol extracts from lamellae contained 14 amino acids, and that all except aspartic acid and alanine caused negative curvatures at 10^{-4} M, as did also $(NH_4)_2SO_4$ and NH_4Cl. No measurements were given. Konishi ascribed hormonal action to these compounds.

Quantitative studies of the effects of various operations on *Agaricus bisporus* (Gruen, 1963) showed that growth ceased within a day when fruitbodies were decapitated before the onset of rapid elongation, and within 3–4 days when they were decapitated at different stages during rapid elongation. The growth rate always decreased without recovery from the first day onwards and the final length was less than normal except when stipes were decapitated close to the end of the growth period. With lamellae alone left on the stipe, even with as little as one-eleventh of the total fresh weight in a pileus, the young stipes attained almost the same final length as with intact pilei. If the central pilear context or a slice of context without lamellae was left on young stipes, residual growth[2] was the same or only slightly greater than after total decapitation. However, the presence of central pilear context during the middle of the stage of rapid elongation caused the stipes to equal the elongation of intact fruitbodies. Thus the stipe will only reach its normal length if lamellae are present until at least the middle of the period of rapid elongation provided that pilear context covers the stipe apex. Without the context, lamellae must remain on the stipe until the end of rapid growth. The central pilear context stops expanding early and could increase residual growth of older stipes by releasing additional amounts of the growth-promoting agent. Diffusate from peripheral young context was inactive (Gruen, 1967), but older context was unfortunately not tested. The presence of pilear context would also enhance acropetal transport of water or nutrients which may be more significant during rapid elongation than at the onset (see also discussion in Gruen, 1963). Expansion of the pilear context itself depends on the presence of lamellae at the periphery, but there is some residual expansion after complete removal of lamellae.

Lamellae left at one side of *Agaricus bisporus* stipes caused only negative curvatures which exceeded 90° in young specimens and averaged only 30° in the middle of the phase of rapid elongation (Gruen, 1967). The above results were indirect but insufficient evidence for production of a growth-promoting agent by the lamellae. Further evidence was obtained by the demonstration that diffusate in plain agar from small portions of lamellae caused negative stipe curvatures after application to the side of young decapitated stipes. In these tests small increases in elongation and consequent small curvatures caused by plain agar were

---

[2]The term residual growth refers to growth which follows removal of all or parts of the pileus.

taken into account. The same diffusate placed symmetrically on the cut stipe apex did not increase growth significantly. Diffusate blocks placed asymmetrically on a few stipes induced a mean curvature of $-47°$ and plain agar blocks $-8°$ (Gruen, unpublished). In contrast to *Flammulina velutipes* (see below) curvature tests by application to the stipe sides of *A. bisporus* seem to be more sensitive than the straight growth method for detecting activity in lamellar diffusate. However, the tests are laborious because each specimen has to be photographed for accurate measurements on the thick stipes which are not parallel sided when young.

F. velutipes can be grown more easily in the laboratory than *A. bisporus*, although its small pilei are less suited for selective removal of different regions (Gruen, 1969). In fruitbodies of *F. velutipes* grown on potato D-glucose agar (PDA) stipe elongation was found to depend on the pileus and specifically on the lamellae. In contrast to *Agaricus bisporus*, less than half the total amount of lamellae in a pileus gave less elongation than intact pilei. Residual growth after decapitation at different stages of development indicated that the stipe depended on the pileus during the whole growth period. The central pilear context left on the stipe apex increased residual growth only slightly during rapid elongation, but when the operation was performed at close to 6 cm length the residual growth almost equaled the growth of intact fruitbodies. Direct dependence of stipe elongation on the lamellae thus seemed to terminate when about three-quarters of the final length had been reached, or somewhat earlier than would be indicated by decapitation experiments. This discrepancy may have been caused by the less rigorous method used for selecting test stipes compared with present procedures (see Culture Conditions and Selection of Test Stipes and Dependence of Stipes on the Pileus and Production of Growth-Promoting Lamellar Diffusate During Rapid Elongation, below). In sharp contrast to *A. bisporus*, neither young nor rapidly elongating *F. velutipes* stipes curved in response to pileus remnants with lamellae left only at one side.

Michalenko (1971) applied pure substances and extracts in lanolin-water emulsion to the side of young decapitated *Flammulina velutipes* stipes close to the apex. The stipes were grown on nutrient-supplemented sawdust, were selected for tests according to a standard method, and were photographed 3 days after application of test substances. Untreated lanolin and ethanol extract from lanolin emulsified with water caused strong negative curvatures but also reduced residual growth. Ethanol-extracted lanolin did not have these effects. Various pure compounds were included in inactive lanolin emulsion at relatively high concentrations because the extent of release to the stipe was unknown. No curvature or growth promotion was observed with seven amino acids, urea, cholesterol, and linoleic acid, all at 0.01 M, or with N-acetyl glucosamine (0.1 M) and casamino acids (2.5%). Several of these compounds actually inhibited residual growth. Glucose (0.1–0.001 M), mannitol (0.01 M), and trehalose (0.1 M) had no effect on curvature or elongation. Various aqueous and ethanol extracts from fresh and lyophilized whole fruitbodies and from pilei and stipes were inactive. Aqueous extracts sometimes caused pronounced swelling at the site of application. Concentrated aqueous diffusate from 6 g of whole fruitbodies (< 5 cm) gave strong negative curvature (mean: $-43°$) and decreased residual growth. Direct application to the stipe of a thick suspension of living hyphal fragments from ground whole fruitbodies (mean length: 5.4 cm) in water or phosphate buffer also gave curvatures averaging $-27°$ and $-34°$, respectively. Purified extracts of total lipids from lyophilized young stipes and from mature pilei and stipes caused significant negative curvature, but lipids from young pilei were inactive. Lipids from young stipes, but not from other portions, reduced residual growth. The results are difficult to interpret because negative curvatures indicate localized growth promotion at the site of application yet they were almost always associated with a decrease in total growth and never with an increase. However, the study showed that stipe curvature can be caused by different extracts, including lanolin extract, which are unlikely to contain the same active compounds. Probably some have nonspecific effects. Also, had the observed curvatures been caused by the growth-promoting agent from the lamellae, activity would have been expected in some extracts which were inactive, such as lipids from young pilei. Lanolin as a carrier had several undesirable features.

A different approach was followed (Gruen, 1976) by testing diffusates in agar from small amounts of *F. velutipes* tissues because operations on the pileus suggested that the active agent was effective at low concentration. The curvature test was abandoned because agar blocks adhere poorly to the stipe side even after blotting, and, like lanolin, are displaced from the elongating apex. A straight-growth test was adopted with agar blocks covering the stipe apex (Figs. 6.1–6.7). Test stipes were grown on nutrient-supplemented sawdust and only the first seven per petri dish from the first crop were used, provided they had a certain minimum diameter. The selection method was based on a detailed study of the gradation of fruitbody dimensions in individual cultures. Lamellar diffusate in plain 1.5% agar increased stipe elongation, but the increase was considerably less than with diffusate from lamellae placed on dilute PDA (= PDA/2)[3]. The discovery that nutrients increased the activity of lamellar diffusate resulted from earlier work on the dependence of stipe growth on the mycelium (see Relationship Between Stipe Elongation and the Mycelium, below). Growth promotion by diffusate was strongest just before and during the beginning of rapid elongation, and declined with increasing fruitbody age. No significant activity was detected at approximately the middle of the phase of rapid elongation. Lamellae on plain agar produced diffusate which showed significant activity only until the onset of rapid elongation. Preliminary experiments indicated that older stipes became insensitive to diffusate from young lamellae, and that older lamellae had no significant effect on young stipes (but see Dependence of Stipes on the Pileus and Production of Growth-Promoting Lamellar Diffusate During Rapid Elongation, below). Very small amounts of young lamellae promoted growth, but there was no direct quantitative relationship within the limits tested. Young pilear context was inactive. *F. velutipes* lamellae thus secrete at least one substance which promotes stipe elongation and the effective concentration is probably quite low.

Forty years after Borriss' studies quantitative work was resumed on the pileus–

---

[3]PDA/2 designates full strength PDA prepared as described by Gruen (1969) and diluted by half with 1.5% plain agar. For one test with PDA/3 (see Effects of Nutrients, below) PDA was diluted 3-fold with the same agar concentration.

Control of Stipe Elongation in Fruitbodies of *F. velutipes* 131

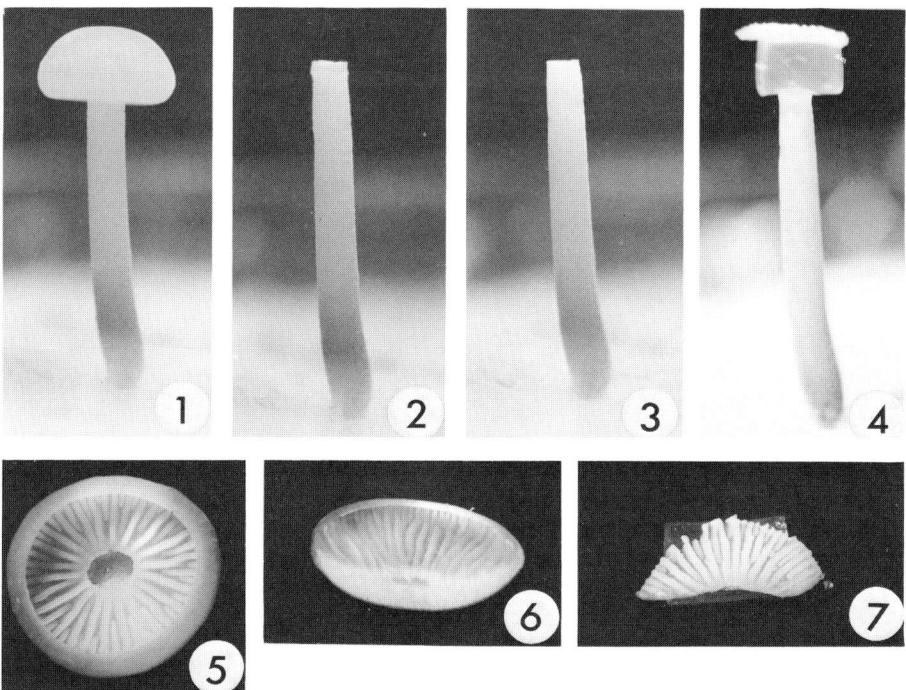

Figs. 6.1–6.7. Steps in the preparation of decapitated *Flammulina velutipes* stipes for testing the growth response to agar blocks with diffusates and other substances. Fig. 6.1 Typical fruitbody of the 3-6 strain in the standard length range (1.5–2.5 cm) used for most tests. Fig. 6.2 Stipe of the same fruitbody after removal of the pileus. Fig. 6.3 Same stipe after cutting a very thin slice (about 1 mm or less) from the apex to obtain a flat surface and to insure removal of any remnant of lamellae. Fig. 6.4 Same stipe with standard 3 × 3 × 5-mm block of 1.5% agar bearing the standard amount of lamellae designated as lamellae/2. For some tests diffusate is obtained in receiver blocks kept separate from the stipe. Fig. 6.5 Detached pileus. Fig. 6.6 Pileus half obtained by cutting tangentially to the opening left by the stipe. Fig. 6.7 Standard agar block with lamellae/2 excised from the pileus portion in Fig. 6.6 by cutting away all context except for a thin supporting layer. The lamellae are placed with the cut surface on the agar.

stipe relationship in species of *Coprinus*. Gooday (1974) excised and decapitated stipes of *C. cinereus* (Schaeff. ex Fr.) S. F. Gray over a wide size range just before and after the onset of rapid elongation (stages IV, V of Borriss, 1934a). He found that they elongated as much as excised whole fruitbodies of the same length both in humid air and with the bases immersed in water. However, only larger specimens (> 35 mm) grew as much as intact fruitbodies attached to mycelium. Although only few measurements were given, excised apical stipe halves elongated as much as intact excised stipes, and excised basal stipe halves did not elongate. Gooday's experiments did not provide evidence for control of stipe elongation by the pileus, but he did not study the effect of decapitation on young pre-elongation stipes (stages II, III) which according to Borriss depend on the pileus. Also, *C. cinereus* fruitbodies do not contain sufficient nutrients for normal elongation until they attain more than one-third of their final length (see below). When younger excised fruitbodies were left intact the absence of continued nutrient supply probably reduced elongation to the level of residual growth of comparable decapitated stipes (see Discussion, below) which did not lose remaining nutrients to the pileus.

Eilers (1974) decapitated excised *Coprinus radiatus* fruitbodies from stage III to the early stage of rapid elongation (V), floated the stipes on water, and found that they elongated less than whole detached fruitbodies. He did not state specifically whether the latter were also floated. Decapitation of stage V fruitbodies at about one-eighth to one-quarter of their final size had no effect on elongation. Stipe elongation was promoted when pilei of increasing age up to approximately the middle of stage V were placed directly on young decapitated stipes (stage III) maintained upright in agar blocks. Eilers concluded from the frequency of certain percentage promotion levels that young pilei were the most effective, but even those close to autolysis caused slight promotion. Eilers regarded this as evidence for production of a diffusible growth regulator by the pileus. When young stipes (stage III) were decapitated and allowed to age before young pilei were placed on the apices, they responded progressively less to the pilei until there was no response when elongation was three-quarters completed. Eilers also reported that short basal stipe portions elongated considerably after removal of the rest of the stipe although they did not do so in the intact fruitbody.

Cox and Niederpruem (1975) reported that vertically bisected, excised primordia of *C. lagopus* sensu Buller (= *C. radiatus*) up to 5 mm length stopped growing. However, similarly treated, somewhat larger primordia reached 20–30 mm. Decapitated excised stipes longer than 5 mm elongated 3- to 4-fold. The number of stipes measured was not given, nor were they compared to whole fruitbodies or to attached specimens.

In stipes of *Coprinus congregatus* Bull. ex Fr. attached to mycelium, Bret (1977a) found that those which were decapitated before and during rapid elongation until close to the end of the growth period grew less than comparable intact fruitbodies. Excised decapitated stipes with their bases in plain agar grew only slightly less than comparable excised whole fruitbodies except in very young specimens. The final length of both was much less than that of attached whole fruitbodies. Attached decapitated stipes grew somewhat more than excised decapitated stipes when the operations were performed from shortly before until shortly after the onset of rapid elongation, but the differences were not compared statistically. Close to the end of the normal growth period, differences between treated and untreated fruitbodies became small and were probably not significant. Bret used only cultures with 2–3 fruitbodies to limit the effect of size variation, and calculated each mean from 40 replicate measurements. His interesting results show that normal stipe growth depends on both the mycelium and the pileus even during rapid elongation. *C. congregatus* thus resembles both *Flammulina velutipes* and *Agaricus bisporus* in regard to the pileus-stipe relationship. Also, excised whole and decapitated fruitbodies of *C. congregatus* grew in a similar manner to those of *C. cinereus* studied by Gooday (see above); however, the latter did not investigate the effect of decapitation on fruitbodies attached to mycelium.

Abnormalities in fruitbody growth offer interesting possibilities for comparative studies of the pileus–stipe relationship. The pilei of several agarics grown in darkness remain rudimentary while the stipes are thin and either abnormally long or stunted. Certain other environmental conditions cause similar effects. Stipes were stunted and usually branched in dark-grown *Lentinus tigrinus* [= *Panus tigrinus* (Bull. ex Fr.) Sing.] (Schwantes and Hagemann, 1965), which lacked pilei, or in *Pleurotus ostreatus* (Jacq. et Fr.) Kummer, which had no or only rudimentary pilei (Gyurkó, 1972). In yellow light *L. tigrinus* produced rudimentary pilei and much longer stipes than those of fruitbodies grown in blue light which were normal. Quantitative comparisons with the length of normal fruitbodies are lacking in the literature but illustrations are often convincing where unusually long or stunted stipes were reported. There is little information on the condition of the lamellae in the rudimentary pilei. Brefeld (1877a, 1877b) noted that lamellae differentiated but did not enlarge in *Coprinus stercorarius* (Bull. ex St-Amans) Fr. which formed long stipes in darkness, and in *C. ephemerus* Fr. which had stunted stipes. He described persistent apical cell divisions in these *C. stercorarius* stipes, and Gräntz (1898) confirmed that their elongation was limited to an apical zone and that rapid elongation was absent. Borriss (1934a, 1934b) did not mention whether or not lamellae were absent in dark-grown *C. lagopus* (= *C. radiatus*) pilei which seem to be fused to the stipe apex. However, the mean stipe length in the dark was far less than for light-grown specimens judging by scattered values given in various graphs (Borriss, 1934a). An apical zone of cell division persisted in the abnormal fruitbodies as long as they grew. The newly formed cells elongated far less than cells of light-grown stipes, and the growth zone remained short and apical. The phase of rapid elongation was absent (Borriss, 1934b). He did not observe cell divisions in illuminated stipes during rapid elongation, and concluded that the type of growth which is normally limited to the differentiation phase continued in the dark-grown stipes throughout their growth period. However, in recent years continued cell divisions have also been found in the elongation phase of light-grown *C. radiatus* (Eilers, 1974) and *F. velutipes* (Wong and Gruen, 1977). They also occurred during rapid elongation in *A. bisporus* where new septa formed in cells of the apical stipe region (Craig et al., 1977). The same pattern probably exists in the other species. This type of cell division may differ from that found in abnormal fruitbodies, but no comparative studies have been made. Reduced cell elongation seems to be one of the main effects of rudimentary lamellae or of the absence of lamellae. Limited cell enlargement also occurs in normal, differentiating primordia (Reijnders, 1963). In neither case is anything known of the physiological interrelations between fruitbody regions. The abnormal growth found in fruitbodies which developed in darkness ceased on exposure to light in *C. lagopus* (= *C. radiatus*) (Borriss, 1934b), *C. congregatus* (Manachère, 1970), and *F. velutipes* (Gruen, unpublished).

Coprinus congregatus fruitbodies grown under continuous illumination at certain temperatures elongated little, and their pilei did not autolyse or form spores (Manachère, 1970; Durand, 1977). Bret (1977b) found that such pilei and aqueous extracts of pilei inhibited residual growth of decapitated stipes which were maintained under normal growth conditions. In this connection morphological mutants of *C. macrorhizus* are of interest because they include an "elongationless" strain with expanding pilei but only very short stipes (Takemaru and Kamada, 1972).

Relationship Between Stipe Elongation and the Mycelium

Studies of interactions between pileus and stipe must consider the role of the mycelium. Dependence of the stipe on the mycelium close to, or after the beginning of rapid elongation has been overlooked repeatedly because stipe elongation, pileus expansion and spore formation occur in detached agaric fruitbodies. Several earlier workers who did not carry out comparative quantitative studies implied or stated specifically that fruitbodies grew normally if they were removed from mycelium during rapid elongation. However, dry weight increases were reported during the 1950s in elongating, attached fruitbodies of *Agaricus bisporus, Flammulina velutipes,* and a species of *Coprinus* and cast doubt on the supposed independence from the mycelium (see Gruen and Wu, 1972a, for a brief review). Obtaining reliable information on the role of the mycelium requires that fruitbody dimensions of detached and attached specimens be compared at different stages of development. Studies of weight changes and requirements for external metabolites would provide additional evidence. Such information is available for only a few species.

F. velutipes fruitbodies were strongly dependent on the mycelium because they grew very little after they were excised and placed horizontally on mycelium or upright with their bases in water. Even disks of mycelium in PDA left attached at the fruitbody base had to be of certain minimum size to support continued stipe growth. Fruitbodies only grew to their normal size when they were excised close to the end of their growth period. The dry weight increase in attached fruitbodies provided evidence for continued inflow of materials during the expansion phase. Various carbohydrates and especially natural extracts with added glucose promoted growth of isolated fruitbodies (Gruen and Wu, 1972a, 1972b). On potato glucose solution the dry weight increase in large elongating fruitbodies paralleled weight losses in the mycelium and stunted fruitbodies, despite continued uptake of glucose from the medium. Materials were probably transported into the large fruitbodies from the rest of the colony because elongation of these fruitbodies was accompanied by degradation of large amounts of glycogen in the mycelium, and of glycogen and cell wall polysaccharides in the stunted fruitbodies. Trehalose, mannitol, and possibly arabitol were probably the main carbohydrates transported into the large fruitbodies (Kitamoto and Gruen, 1976). Elongation of the large fruitbodies was also accompanied by an increase in total amounts of organic nitrogen in the stipe and pileus, and by nitrogen loss from the mycelium (Wong, 1978).

Excised *Coprinus cinereus* fruitbodies elongated, but they had to reach more than one-third of their final length prior to excision before they elongated as much as attached fruitbodies (Gooday, 1974). Nutrients increased elongation of small excised fruitbodies. Since they were excised at the beginning or during rapid elongation it can be concluded that their normal growth depended on continued connection with the mycelium during at least the early part of that phase.

Bret's (1977a) results demonstrated that even during rapid elongation *Coprinus congregatus* fruitbodies only attained their normal length if they remained attached to mycelium. According to Robert (1977) the dry weight of normal *C. congregatus* fruitbodies grown in alternating darkness and light increased throughout the growth period, although less during than before rapid elongation. Carbohydrate concentration reached a peak at the beginning of elongation, and the concentration of the alcohol-soluble fraction decreased thereafter. The decrease in absolute amounts occurred mainly in the pilei which autolyze. In light-induced fruitbodies which elongated in darkness the pilei did not expand or autolyze and formed few spores. The dry weight of these stipes was greater than in normal fruitbodies, and alcohol-soluble carbohydrates and proteins increased until growth ceased. These results demonstrate that materials other than water are translocated into elongating stipes of this *Coprinus*.

Turner (1977) reported that *Agaricus bisporus* fruitbodies excised at the beginning of rapid elongation continued to grow at about half the rate of attached fruitbodies, and that the pileus diameter was smaller. The limited elongation of the excised specimens suggests that normal elongation depends on the mycelium. However, the observation periods of 48 h for excised and 24 h for attached fruitbodies were very short considering the long growth period of this species (Hagimoto and Konishi, 1959; Gruen, 1963).

Stipes must be decapitated or at least substantial portions of the pileus must be removed to test whether extracts or pure compounds can replace the role of lamellae in elongation. These operations could interfere with translocation of materials from the mycelium which are required for normal stipe elongation. Very little information is available on this point. Hagimoto (1964) found that residual growth of *A. bisporus* stipes decapitated soon after the onset of rapid elongation (32 mm) was accompanied for 3 days by fresh and dry weight increases which were much smaller after the first day than in comparable pileate stipes. At a slightly later stage (42 mm) dry and fresh weight increases in decapitated and intact stipes were the same, but the test period was only 2 days. Kamada et al. (1976) decapitated wild type *C. macrorhizus* fruitbodies at approximately the beginning of rapid elongation and found that the content of protein and RNA, but not DNA, increased more than in intact fruitbodies. Residual growth was not mentioned but presumably took place. The limited information available indicates that at least some materials continue to be translocated from the mycelium into decapitated stipes.

Plant Growth Regulators and Nucleotides in Relation to Fruitbody Growth

Many years ago the *Avena* curvature test was used to demonstrate auxin activity in extracts from fruitbodies of *Boletus edulis* Bull. ex Fr. (Nielsen, 1932; Almoslechner, 1934), and from the pileus and stipe of *Agaricus bisporus* (Bouillenne-Walrand et al., 1953). Urayama (1956) mentioned that diffusate from *A. bisporus* lamellae was active in the *Avena* test. Indoleacetic acid (IAA) was later detected chromatographically in stipes and pilei of *A. bisporus* (Konishi and Hagimoto, 1961). Very low auxin activity was found in pilei and stipes of *Lentinus tigrinus* (Rypácek and Sladký, 1972, 1973) with the *Avena* coleoptile section bioassay. Gibberellin-like and cytokinin-like activities were also detected. Pegg (1973) found significant levels of gibberellin-like substances in the lamellae and pilear context of *A. bisporus,* but not in the stipe. Cytokinin-like activity was reported in extracts from fruitbodies of *Coprinus micaceus* Fr. (Szabó et al., 1970).

Because IAA and other growth hormones of vascular plants are also synthesized by fungi many efforts have been made to determine whether these compounds play a similar role in fungi themselves, including in several Agaricales. IAA, synthetic auxins, and auxin antagonists have been added to culture media but have usually been inhibitory at high concentration, or ineffective, or caused weak promotions of mycelial growth of doubtful significance (see review in Gruen, 1959). Increased dry weight yield of *Flammulina velutipes* fruitbodies was reported with IAA under some conditions (Aschan-Åberg, 1958). Changes in auxin, gibberellin, and cytokinin levels in mycelium, pilei, and stipes of *L. tigrinus* at different stages of development were interpreted by Rypáček and Sladký (1972, 1973) as evidence for a growth regulating role of these compounds. Sladký and Tichý (1974) added IAA, gibberellic acid, and kinetin separately to mycelium at the time of primordium formation, and with some concentrations observed significant increases in number, weight, or size of fruitbodies. The amounts of endogenous growth regulators also increased. High concentrations of IAA increased the number of fruitbodies and decreased the size on malt extract–cellulose medium. This effect on size may have been caused by competition for nutrients among the fruitbodies. None of these results can be regarded as proof for a hormonal role of these compounds in the fungus.

Jeffreys and Greulach (1956) applied solutions of IAA, NAA, and 2,3,5-triiodobenzoic acid directly to decapitated *Coprinus sterquilinus* stipes but low concentrations did not promote growth and high concentrations were inhibitory (+ curvature). Unilateral application of IAA in agar to pileus remnants without lamellae did not induce stipe curvature in *A. bisporus* (Urayama, 1956; Konishi and Hagimoto, 1961). At 10^{-3} M, but not at 10^{-2} M, IAA in lanolin gave significant small negative curvatures in decapitated *F. velutipes* stipes (Michalenko, 1971), but the problem of interpreting curvatures obtained with this method have been noted above. IAA (10^{-3} M) with glucose fed through the base of whole excised *F. velutipes* fruitbodies had no effect on elongation (Gruen and Wu, 1972b). Different concentrations of gibberellins A_1, A_3, and A_5 applied directly to young pilei of *A. bisporus* had no effect on growth except that a high concentration of gibberellic acid inhibited pileus and stipe growth (Pegg, 1973).

Fruiting could be induced on monokaryotic mycelium of *Coprinus macrorhizus* f. *microsporus* by low concentrations of adenosine-3′,5′-cyclic monophosphate (cAMP) (Uno and Ishikawa, 1971, 1973b), and also by adenosine-3′-monophosphate (3′-AMP) and theophylline (Uno and Ishikawa, 1973a). The latter compounds probably stimulated fruiting because they inhibit phosphodiesterase-mediated breakdown of cAMP to 5′-AMP. Extracts from fruitbodies of other agarics, including *F. velutipes,* induced fruiting in monokaryotic *C. macrorhizus* (Uno and Ishikawa, 1971), but the presence of cAMP in these extracts has not been demonstrated. Matthews and Niederpruem (1972) mentioned that cAMP accelerated formation of hyphal aggregates which initiate dikaryotic fruiting in *Coprinus lagopus* (= *C. radiatus*). Whether cAMP plays a role in fruitbody elongation has not been reported (see Ineffectiveness of Plant Growth Regulators and Nucleotides in Elongation of Decapitate Stipes, below) although Bu'Lock (1976) mentioned that preliminary work on changes in cAMP level suggested a role in development of *C. lagopus* (see also Chapter 5).

Recent Studies on *Flammulina velutipes*

Culture Conditions and Selection of Test Stipes

Large-scale testing of growth-promoting substances on agaric fruitbodies requires a continuous supply of fruitbodies of adequate and comparable size. Several fruitbodies per culture must be used to avoid the need for maintaining excessive numbers of cultures. Clustering should be minimized because individual fruitbodies must be manipulated. For the work on *F. velutipes* the best fruitbodies in terms of numbers, size, pileus shape, and adequate dispersion on the mycelium are obtained in the light on sawdust–wheat bran–malt extract medium in 9 × 2-cm petri dishes (Gruen, 1976). Although a synthetic medium would have been preferable, the 3-6 strain of *F. velutipes* used in this work produced only small thin fruitbodies on several media tested. These were unsuited for work on the pileus-stipe relationship and for studies of the distribution and translocation of endogenous carbohydrates and nitrogen compounds. Among media which proved inadequate was Plunkett's (1953) medium as modified by Naudy-de Serres et al. (1975). In recent tests of this liquid medium the 3-6 strain formed many small fruitbodies, few of which attained a maximum final length of 2–3 cm (Wong, 1978). The mean length appears to be roughly comparable to the 1 cm mentioned by the French workers for their strain grown in light, and is only about one-tenth the size obtained on sawdust. Fewer and somewhat smaller fruitbodies were produced on potato glucose agar or solution than on sawdust, and the pilei were often irregular in shape. On the solution densely clustered, aborted, and stunted individuals accompanied the large specimens. The number of primordia varies on sawdust medium, but the mean number of fruitbodies of over 2 cm final length is close to 12 per culture in the first crop, which was used exclusively in the present work. The fruitbodies grow singly or in pairs with their bases close together or sometimes fused (Fig. 6.8). Clusters of three or more elongating specimens are less common and often only one or two individuals in a cluster will fall within the number used (see below). There is always a size gradation in each culture (Figs. 6.8, 6.9). It is reflected in the length, pileus diameter, and stipe diameter (Gruen, 1976), and must be taken into account in growth studies.

All fruitbodies used for the present work were ranked according to decreasing size when they were less than about 1 cm long. Among individuals of the same length the thickest was ranked higher. Previously the 7 largest were used, but the

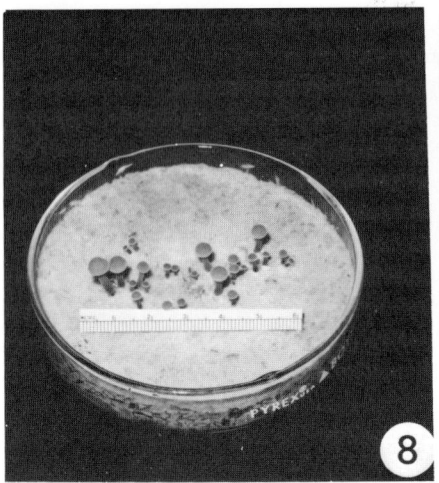 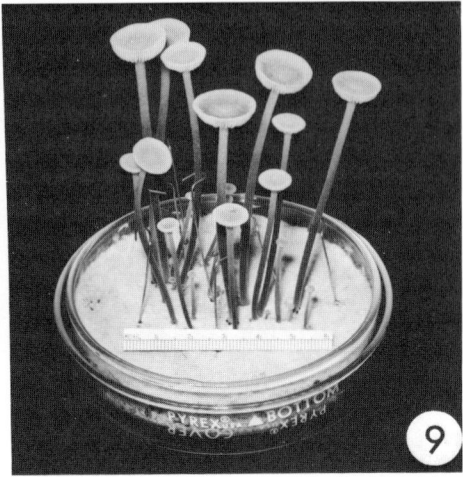

Figs. 6.8–6.9. Petri dish culture of *Flammulina velutipes*, strain 3-6, on nutrient-supplemented sawdust medium. Fig. 6.8 Appearance at 31 days after inoculation when fruitbodies are ranked 1–5 by decreasing size. Excess fruitbodies are removed at this time from cultures used for tests. Fig. 6.9 Appearance 4 days later when fruitbodies had attained between one-half and three-quarters of their final length. At advanced stages of development fruitbodies are supported by hooked Chromel-A wires when their growth is followed to completion.

number was reduced to 5 in recent work on stipe growth. Although fruitbody initiation cannot be determined precisely by gross observation, the largest fruitbodies are probably also the earliest to be initiated. The earlier description of culture methods (1976) inadvertently omitted mentioning that all fruitbodies in excess of the number used were removed.[4] A further selection criterion consisted of excluding stipes of less than 1.5 mm apical diameter from those which were decapitated for tests because thin fruitbodies did not even attain 5 cm final length. Most such thin stipes fell into ranks 6 and 7, and this was another reason for selecting the 5 largest. Only few of these had to be eliminated at the 1.5–2.5-cm stage which was chosen as the standard size for most tests, mainly because it corresponds to the beginning of rapid elongation when the response to lamellar diffusate is maximum. The final length of intact fruitbodies averaged 11.1 cm for rank 1 and 8.5 cm for rank 5. These means compare with 11.6 and 8.0 cm, respectively, reported previously for cultures where all fruitbodies were allowed to elongate. Thus, the maximum difference in final length among ranks was 1 cm more in the earlier study, probably because of competition among larger numbers of elongating fruitbodies.

To assess the effect of variation, growth promotion caused by lamellar diffusate in PDA/2 was averaged for decapitated young stipes from each of the 5 ranks.

---

[4] Excess fruitbodies were not removed in the comparisons between early and final ranking of intact fruitbodies by size (Gruen, 1976, Fig. 1). This was to take into account individuals which elongated more than those that were included initially among the 7 largest.

For every rank the promotion was calculated from the means of 15–27 stipes each treated with the nutrient alone and with diffusate. All data were obtained from cultures inoculated on the same dates and from stipes of all ranks for both treatments. Seven culture series had to be combined to obtain reasonably large samples (total 215), and the number of measurements differed for the various series and ranks. Promotions ranged from 1.28 to 1.89 cm, but the means for 4 of the 5 ranks differed little (1.28–1.46 cm). The high value for rank 4 did not differ statistically from the lowest promotion, and included a large proportion of measurements from culture series with strongly responding stipes. The response to other treatments also did not differ significantly among decapitated stipes from the 5 ranks.

Although standardized growth conditions and selection methods reduced the variation among test stipes it still remained considerable. Fruitbodies which responded much less than others often occurred in cultures on the same batch of medium and on mycelium derived from inoculum of the same age taken from near the edge of one stock colony. To minimize the effect of variation, control blocks and blocks with various test compounds were applied to stipes of all ranks taken as far as possible from culture series inoculated on the same dates. Also, comparisons among different substances or concentrations were made as far as possible with stipes from the same series or at least from consecutive series.

Dependence of Stipes on the Pileus and Production of Growth-Promoting Lamellar Diffusate During Rapid Elongation

Earlier work suggested that growth of older stipes depended on the pileus even when growth-promoting diffusate could no longer be detected in the lamellae. This discrepancy prompted a re-examination of the upper limit of dependence of stipe elongation on the pileus. The period of rapid elongation extended to about 8 cm and the mean final length was 9.7 cm in intact fruitbodies selected as described above. Fruitbodies were decapitated as shown in Figs. 6.1–6.3. Stipes were also deprived of lamellae but left with the central pilear context, or were left only with bilateral groups of lamellae held together by a very thin layer of context. These operations were described previously for *Agaricus bisporus* and *Flammulina velutipes* (Gruen, 1963, 1969). About one-third of all lamellae in a pileus were retained on the stipes. Stipes which were decapitated or left only with pilear context at 4–7.9 cm fruitbody length grew significantly less than intact specimens (Table 6.1). In the same range lamellae alone increased elongation significantly over that of decapitated stipes although injury to the apex was the same in both operations. The lamellae became less active with advancing age and neither lamellae nor whole pilei influenced stipe growth at over 8 cm. Growth with the reduced number of lamellae was significantly less than with whole pilei up to 6–6.9 cm, but averaged nevertheless about three-quarters of the elongation of intact fruitbodies. Growth with the lamellae alone also exceeded that of stipes with the central pilear context, but significantly only up to the 4–4.9 cm range. This is lower than the limit for promotion over growth of decapitated stipes (7–7.9 cm) because the context itself increased residual growth slightly. The discrepancy does

Table 6.1. Residual Elongation of Decapitated Stipes, and of Stipes with Central Pilear Context or with Lamellae Alone Compared to Intact *Flammulina velutipes* Fruitbodies During and After Rapid Elongation.

Length at time of operation (cm)	Elongation			
	Decapitated stipes (cm)	Stipes with central pilear context (cm)	Stipes with lamellae alone (cm)	Intact fruitbodies[a]
4.0–4.9	2.3[b]	2.6[b]	4.3[b,c,d]	5.2
5.0–5.9	2.2[b]	2.7[b,c]	3.0[b,c]	4.3
6.0–6.9	1.9[b]	2.1[b,c]	2.3[b,c]	3.3
7.0–7.9	1.6[b]	1.8[b]	2.2[c]	2.4
8.0–8.9	1.5	1.4	1.1	1.5
9.0–9.9	1.1	1.2	—	1.0
10.0–10.9	0.6	—		0.4

Number of stipes measured per length range: decapitated stipes, 20–39; stipes with central pilear context, 18–29; stipes with lamellae alone, 19–36; intact fruitbodies, 90–192.
[a]Differences between the midpoint of each length range and the mean final length of all fruitbodies which equalled or exceeded the lower limit of each range. Mean final length of all intact fruitbodies, 9.7 cm.
[b]Significantly less than the corresponding mean for intact fruitbodies at $P = 0.05$.
[c]Significantly greater than the corresponding mean for decapitated stipes at $P = 0.05$.
[d]Significantly greater than the corresponding mean for stipes with pilear context at $P = 0.05$.

not signify that elongation can proceed normally without lamellae at over 5 cm because it can be assumed that the stipes would have grown more if most lamellae could have been retained while the context was removed. Also tests with slices of context alone at 5–5.9 cm yielded even less residual growth than decapitated stipes or stipes covered only by context from the center of the pileus. It can be concluded that stipe elongation continued to depend on the pileus, and specifically on the lamellae, during the entire phase of rapid elongation up to about 8 cm fruitbody length. The stipe then became independent during the final period of declining growth.

Previously it had been found (Gruen, 1976) that lamellar diffusate from older fruitbodies (5.1–6.0 cm) grown on one batch of sawdust caused slight, but significant, growth promotion in stipes of the same age while the same tests gave negative results on a second batch. Also, preliminary tests with few stipes from the latter medium had shown that diffusate from old lamellae had no significant effect on young stipes, or young lamellae on old stipes. The trend of the response curves on both media indicated that the upper limit of detectability for active diffusate fell between 4 and 6 cm fruitbody length. This limit was re-examined with larger numbers of fruitbodies grown on sawdust similar to the second batch used previously. Diffusates in PDA/2 from lamellae of 5.0–5.9 and 6.0–6.9 cm fruitbodies were applied to stipes of the same ages and to young stipes. Diffusate from young lamellae was also tested on the older stipes. Because old lamellae are much larger than young ones, quarter portions were used for most tests instead of the standard

Control of Stipe Elongation in Fruitbodies of F. velutipes 141

halves shown in Fig. 6.7. Diffusate from older lamellae of both length ranges caused small but significant growth promotion in stipes of the same age (Table 6.2). Also, elongation of young stipes was increased significantly by lamellae from 6.0–6.9 cm fruitbodies but not from the younger 5.0–5.9-cm stage. This discrepancy was probably caused by variation in the material, and by the fact that the activity was at the lower limit of sensitivity of the method. Diffusate from young lamellae promoted growth in older stipes of both length ranges. Thus lamellae from 6–7-cm fruitbodies still released active diffusate, and older stipes remained sensitive to the lamellar growth factor. At 6–7 cm approximately two-thirds of the mean final length had been reached. Although the results from surgical experiments indicate that even older lamellae produce the growth-promoting factor (Table 6.1), the diffusate from small amounts of such lamellae was inactive. Probably much larger amounts will have to be tested to detect significant activity.

Elongation caused in young or older stipes by diffusate from small amounts of lamellae did not approach that of intact fruitbodies. Young test stipes in the stan-

Table 6.2. Tests for the Presence of Growth-Promoting Activity in Diffusates from Lamellae, Pilear Context, and Stipe of *Flammulina velutipes* at Different Stages of Development.

Source of diffusate	Length ranges (cm)		Elongation as increase over controls without diffusate (cm)	
	Donor	Test stipes		
Lamellae[a]	5.0–5.9	5.0–5.9	0.45[d]	(21, 23)
Lamellae	1.5–2.5	5.0–5.9	0.63[d]	(23, 23)
Lamellae	5.0–5.9	1.5–2.5	0.14	(24, 48)
Lamellae	1.5–2.5	1.5–2.5	1.22[d]	(45, 48)
Lamellae	6.0–6.9	6.0–6.9	0.43[d]	(20, 23)
Lamellae	1.5–2.5	6.0–6.9	0.46[d]	(20, 23)
Lamellae	6.0–6.9	1.5–2.5	0.27[d]	(25, 25)
Lamellae	1.5–2.5	1.5–2.5	1.23[d]	(27, 25)
Pilear context[b]	5.0–5.9	5.0–5.9	0	(27, 17)
Stipe segments[c]				
apical	5.0–5.9	5.0–5.9	0.08	(27, 23)
middle	5.0–5.9	5.0–5.9	−0.06	(34, 17)
basal	5.0–5.9	5.0–5.9	−0.23	(21, 17)
apical	1.5–2.5	1.5–2.5	0.03	(28, 16)
basal	1.5–2.5	1.5–2.5	0.14	(12, 9)

Diffusates in dilute potato glucose agar (PDA/2) left on test stipes for 3 days. Controls, PDA/2 alone. In parentheses, number of stipes with diffusate followed by number of control stipes.
[a]Lamellae/2 portions (see Fig. 6.7) from the 1.5–2.5-cm range, and quarter portions from 5–6.9-cm fruitbodies.
[b]Pilear context from slightly less than half the pileus placed with the lower surface on receiver blocks.
[c]Five-mm segments cut from the apical, middle, and basal portions of the respective regions; juncture with mycelium cut from basal segments; all segments placed in normal orientation on receiver blocks.
[d]Significantly different from controls at $P = 0.05$.

dard length range only attained a mean final length of 4.3 cm with the diffusate, or only 44% of the normal length. Some of the factors which limit the response to diffusate from excised lamellae are discussed below, especially in Some Characteristics of the Lamellar Diffusate.

Diffusates from Fruitbody Regions other than Lamellae

Diffusate from the context (trama) of young pilei was inactive in the stipe growth test (Gruen, 1976). Further studies (Table 6.2) also showed that context from older pilei applied on PDA/2 was inactive on stipes of the same age (see also Gruen, 1979). Nor was any activity detected in 5 mm segments from apical, middle, and basal regions of older stipes, or from upper and basal halves of young stipes. Limited elongation of the upper stipe segments occurred on the receiver blocks. These tests confirmed that only lamellae release detectable amounts of growth-promoting substance.

Effect of Nutrients

Dependence of fruitbody elongation on external nutrients (Relationship Between Stipe Elongation and the Mycelium, above) had suggested that the growth-promoting action of lamellae might be influenced by nutrients. Diffusate from young excised lamellae placed on PDA/2 increased growth of decapitated stipes more than diffusate in plain agar. Because it was unknown which endogenous metabolites in fruitbodies influence the lamellae, tests of various nutrients were based on the response of whole excised fruitbodies (Gruen and Wu, 1972b). However, lower concentrations were used than for whole fruitbodies. The following were applied in agar to decapitated stipes (1.5–2.5 cm) with and without lamellae: vitamin-free casein hydrolysate (Difco Casitone), 0.005, 0.01, and 0.02% with and without 0.05 M glucose, and 0.05% without glucose; yeast extract, 0.005 and 0.02% with and without glucose; glucose 0.005, 0.05, and 0.1 M; asparagine 10^{-2}, 10^{-3}, and 10^{-4} M; asparagine 10^{-2}, 5×10^{-3}, and 10^{-3} M with glucose 0.05 M; PDA/2; PDA/3; and dilute potato extract agar (PA/2). Controls and diffusates were tested for each nutrient on stipes from the same series of cultures but only some of the different nutrients could be compared in parallel cultures. The lamellae placed on blocks containing any of the nutrients tested increased elongation significantly compared to controls with nutrient alone, but the promotion was significantly greater than on plain agar only with PDA/2, PDA/3, PA/2, and barely with 0.02% casein hydrolysate. In groups of culture series inoculated on different dates over a 14-month period, the promotion with lamellae on plain agar ranged from 34 to 55%[5] (mean: 43%) and for PDA/2 from 60 to 103% (mean: 78%). PA/2 caused less promotion than PDA/2 in stipes from parallel cultures, but the

---

[5]Calculated as the difference between percent promotions over initial length for controls and diffusate.

difference was not significant at $P = 0.05$. Earlier tests of potato extract without added glucose (Gruen, 1976) had only indicated promotion equal to diffusate in plain agar but smaller blocks were used and the tests were not performed in parallel cultures. The extract may have contained some glucose from hydrolysed starch, but glucose did not significantly increase the effect of lamellae compared to plain agar or when added to casein hydrolysate, yeast extract, or asparagine.

Without lamellar diffusate, PDA/2 and casein hydrolysate (0.05%) inhibited residual growth slightly but significantly compared to plain agar, while glucose alone (0.05 M) and yeast extract with glucose caused significant small increases. The chitin precursor N-acetyl-D-glucosamine was tested at 10^{-2} and 10^{-3} M on decapitated stipes but had no effect on elongation compared to plain agar.

In tests of the effects of nutrients on production of lamellar diffusate the agar blocks contained both diffusate and nutrients. This was also the case in various other experiments whether lamellae on receiver blocks were placed on the stipe at the time of decapitation or kept separately. It is possible that nutrients such as PDA which increase the response to lamellar diffusate do not act on the lamellae but only on the stipe in conjunction with diffusate. To separate nutrient effects on the two regions another test procedure was devised. Agar rings, 5 mm in diameter (Fig. 6.10) can be slipped easily over decapitated stipes of *Flammulina velutipes* and adhere closely to the surface. However, they are gradually displaced from the elongating apex. Lamellae on plain agar were applied on the apex, and either PDA/2 or plain agar rings were placed around the stipe close to the apex (Fig.

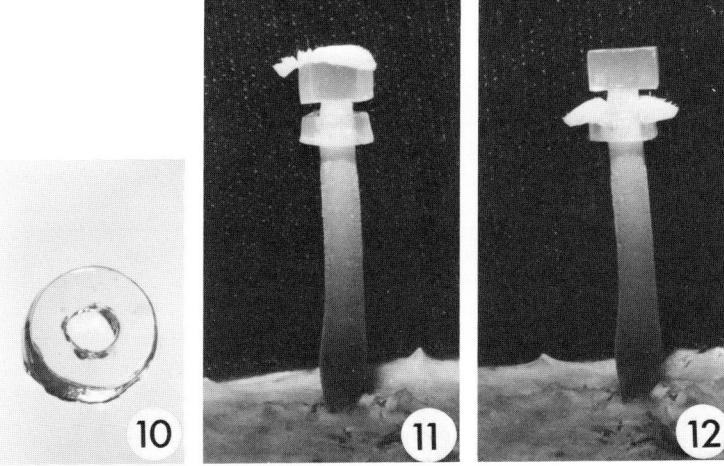

Figs. 6.10–6.12. Use of agar rings for separate application of diffusates and other substances to the apex and lateral surface of decapitated *Flammulina velutipes* stipes. Fig. 6.10 Ring of 1.5% agar, 5 mm in diameter and 3 mm thick, perforated with a glass capillary to fit over stipes of 2 ± 0.5 mm diameter. Fig. 6.11 Decapitated stipe (2.3 cm) with ring and with lamellae/2 portion (see Fig. 6.7) placed on $3 \times 3 \times 5$-mm agar block applied on the apex. Fig. 6.12 Decapitated stipe with agar block on apex and with 2 lamellae/2 portions placed on ring at opposite sides of the stipe. Initially, the lamellae are not in contact with the stipe surface.

6.11). Growth promotion did not differ significantly between the two treatments (Table 6.3), and was the same (32–44% relative to initial lengths) as for lamellar diffusate in plain agar without rings. These results suggest that the growth promotion which always occurred with lamellae on apically applied PDA/2 blocks is caused primarily by the effect on the lamellae (see Discussion, below). However, the evidence is not yet conclusive because the extent of penetration of PDA/2 from the ring into the stipe is unknown. Such penetration is suggested by the fact that PDA/2 rings reduced residual growth in two of three test series (Table 6.3) as occurred also with apically applied PDA/2 without diffusate. Single standard portions of lamellae (Fig. 6.7) placed on PDA/2 or plain agar rings had no effect on elongation. Two portions of lamellae on rings (Fig. 6.12) promoted growth significantly, but the promotion was the same with both types of agar, and even less than with single portions of lamellae on apical blocks of plain agar (Table 6.3). Penetration of the active diffusate through the stipe sides seems to be limited. Further studies are being carried out with this system, especially with lower agar concentration in the ring.

A drop of the dye light green (0.1%) was placed on agar blocks applied to the apex of decapitated stipes in the usual manner. The dye moved through the block and reached approximately 0.7–1 cm below the apex in 2–3 h. Growth continued. The movement occurred in all stipe portions and seemed to be mainly interhyphal because most of the dye was gradually released on immersion of stipe slices in water. By contrast, light green applied to plain agar rings diffused through the

Table 6.3. Elongation of Decapitated *Flammulina velutipes* Stipes After Simultaneous Application of Lamellar Diffusate to the Apex and Potato Glucose Agar Rings to the Stipe Sides (see Fig. 6.11), and After Application of Lamellar Diffusate in Agar Rings (see Fig. 6.12).

Agar type		Elongation (cm)		
Apical block	Ring	No lamellae	Lamellae	Promotion
Lamellae/2[a] on apical block				
plain	PDA/2	1.03 ± 0.15	1.60 ± 0.37	0.57[b,c]
plain	plain	1.30 ± 0.17	2.09 ± 0.35	0.79[c]
Lamellae/2 on ring				
plain	PDA/2	1.25 ± 0.20	1.43 ± 0.19	0.18[b]
plain	plain	1.65 ± 0.37	1.61 ± 0.28	−0.04
2 Lamellae/2 on ring				
plain	PDA/2	1.18 ± 0.23	1.59 ± 0.28	0.41[b,c]
plain	plain	1.22 ± 0.23	1.58 ± 0.18	0.36[c]

Initial fruitbody length, 1.5–2.5 cm. Growth period, 3 days. Number of stipes per treatment: without lamellae, 17–28; with lamellae, 22–27. Variation as ± 95% confidence limits.
[a] Standard amount of lamellae excised from pileus halves (see Fig. 6.7).
[b] Not significantly different at $P = 0.05$ from promotion with ring of plain agar.
[c] Significant at $P = 0.05$.

agar but stained the stipe surface only slightly and moved very little beyond the rings even after 2–3 days. On the other hand a drop of the dye placed on the stipe surface in the growth zone penetrated rapidly, and after 2 h the dye had moved several millimeters longitudinally in both directions but not to the opposite side. Although movement of light green (MW = 792.8) and of metabolites is not necessarily the same, the experiments with the dye and with lamellar diffusate suggest that diffusion from agar is rapid into the cut stipe apex but is probably a limiting factor in testing compounds by application on the stipe sides.

Ineffectiveness of Plant Growth Regulators and Nucleotides in Elongation of Decapitate Stipes

IAA, gibberellic acid, 6-benzylaminopurine, cAMP, 3'-AMP, and 5'-AMP were applied in plain purified agar to decapitated stipes in the usual manner to determine whether they could replace the growth-promoting effect of lamellae. Glucose was added to the agar in some experiments. Nucleotides and IAA were also tested together with lamellae. Theophylline was tested with and without lamellae for possible interference with endogenous cAMP breakdown. Test and control blocks were applied to stipes from the same series of cultures, and different concentrations of most compounds were tested in culture series with the same or overlapping inoculation dates.

Table 6.4 shows that neither growth hormones, nucleotides, nor theophylline influenced stipe elongation except for inhibition by high concentrations of IAA and 6-benzylaminopurine. Addition of glucose to IAA, benzylaminopurine, and cAMP had no effect on growth. Similarly, growth promotion by diffusate from lamellae exposed to the nucleotides and theophylline did not differ significantly from the response to diffusate in plain agar (Table 6.5). Lamellae treated with IAA gave slightly but significantly less elongation than on plain agar.

It is unknown whether the plant growth hormones and nucleotides tested are normally present in fruitbodies, but the consistent negative results are strong evidence against their role as growth regulators in *Flammulina velutipes* (see also Plant Growth Regulators and Nucleotides in Relation to Fruitbody Growth, above). This conclusion is reinforced by the fact that the compounds were tested after the known endogenous source of a growth-promoting agent was removed, and at a stage of development when the stipes were most sensitive to lamellar diffusate.

Some Characteristics of the Lamellar Diffusate

PDA/2 strips equal in size to four standard blocks were covered by pre-soaked strips of dialysis membrane (pore size 4.8 nm, retains 12,000 MW) which were slightly longer and wider than the agar. Four lamellae portions (Fig. 6.7) were placed on each strip. After 24 h the agar was cut into standard blocks and tested

Table 6.4. Tests of Plant Growth Regulators, Nucleotides, and Theophylline on Elongation of Decapitated *Flammulina velutipes* Stipes.

	Elongation (cm)	
Test compounds in agar blocks(M)	Test compound	Difference from controls[a]
IAA 10^{-5}	1.13 ± 0.25 (24)	0.14
IAA 10^{-4}	1.64 ± 0.18 (32)	0.24
IAA 10^{-3}	0.68 ± 0.15 (20)	−0.41[b]
IAA 10^{-5} + glucose 0.05	1.32 ± 0.24 (21)	0.25
IAA 10^{-4} + glucose 0.05	1.15 ± 0.17 (25)	0.08
Gibberellic acid 10^{-5}	1.32 ± 0.31 (27)	0.16
Gibberellic acid 10^{-3}	1.22 ± 0.16 (28)	0.06
6-Benzylaminopurine 10^{-7}	1.07 ± 0.19 (24)	0.01
6-Benzylaminopurine 10^{-5}	1.06 ± 0.17 (25)	−0.26[b]
6-Benzylaminopurine 10^{-3}	0.44 ± 0.06 (26)	−0.88[b]
6-Benzylaminopurine 10^{-5} + glucose 0.05	1.23 ± 0.34 (14)	−0.20
cAMP 10^{-6}	1.45 ± 0.39 (18)	−0.04
cAMP 10^{-5}	1.28 ± 0.23 (37)	−0.03
cAMP 10^{-4}	1.75 ± 0.38 (18)	0.26
cAMP 10^{-5} + glucose 0.05	1.69 ± 0.25 (21)	−0.29
cAMP 10^{-5} + theophylline 10^{-6}	0.94 ± 0.17 (20)	−0.24
cAMP 10^{-5} + theophylline 10^{-5}	1.41 ± 0.21 (31)	0.12
Theophylline 10^{-6}	1.06 ± 0.29 (22)	0.02
Theophylline 10^{-5}	1.54 ± 0.33 (17)	0.23
3'-AMP 10^{-5}	1.09 ± 0.29 (20)	−0.09
5'-AMP 10^{-5}	1.07 ± 0.34 (14)	−0.11

Initial fruitbody length, 1.5–2.5 cm. Growth period, 3 days. Number of stipes with test compounds in parentheses. Variation as ± 95% confidence limits. IAA = indole-3-acetic acid; cAMP = adenosine-3',5'-cyclic monophosphate, Na salt; 3'-AMP = adenosine-3'-monophosphate, Na salt; 5'-AMP = adenosine-5'-monophosphate, Na salt.

[a] Control stipes with plain agar or with 0.05 M glucose where the latter was added to the test compound. Number of control stipes with each test compound and concentration, 12–49 (mean: 26).
[b] Difference significant at $P = 0.05$; negative if elongation with test compounds less than controls.

on decapitated stipes. Controls consisted of blocks cut from PDA/2 strips which had been covered for 24 h with dialysis membrane without lamellae. The results in Table 6.6 show that the active material diffused through the membrane.

Lamellar diffusate was collected in PDA/2 strips for 24 h, the agar was boiled 30 min, made up to the initial volume, cooled, cut into standard blocks, and applied to decapitated stipes. For comparison, diffusate obtained in the same manner was merely remelted and made into blocks, and PDA/2 was boiled and treated in the same manner as the boiled diffusate. The results of growth tests (Table 6.6) showed that the active agent in lamellar diffusate was heat stable. Unheated 24-h diffusate in PDA/2 strips was not tested at the same time, but

Table 6.5. Tests of IAA, Nucleotides, and Theophylline on the Response of Decapitated *Flammulina velutipes* Stipes to Excised Lamellae.

Test compounds in agar blocks (M)	Elongation (cm)		
	Lamellae on test compounds	Promotion over test compounds alone	Promotion by lamellae on plain agar
IAA 10^{-4}	1.49 ± 0.29 (42)	−0.15	0.39[a]
cAMP 10^{-6}	2.55 + 0.66 (16)	1.10[a]	0.98[a]
cAMP 10^{-5}	2.23 ± 0.35 (38)	0.95[a]	0.76[a]
cAMP 10^{-4}	2.29 ± 0.53 (16)	0.54	0.98[a]
cAMP 10^{-5} + glucose 0.05	3.02 ± 0.43 (21)	1.33[a]	0.80[a,b]
cAMP 10^{-5} + theophylline 10^{-6}	1.67 ± 0.38 (19)	0.73[a]	0.67[a]
cAMP 10^{-5} + theophylline 10^{-5}	2.60 ± 0.61 (18)	1.17[a]	1.02[a]
Theophylline 10^{-6}	1.77 ± 0.36 (31)	0.71[a]	0.37[a]
Theophylline 10^{-5}	2.27 ± 0.42 (18)	0.73[a]	0.88[a]
3′-AMP 10^{-5}	1.65 ± 0.29 (21)	0.56[a]	0.67[a]
5′-AMP 10^{-5}	1.51 ± 0.36 (19)	0.44[a]	0.67[a]

Initial fruitbody length, 1.5–2.5 cm. Growth period, 3 days. For each compound and concentration, number of stipes with lamellae in parentheses, without lamellae, 14–37 (mean: 22), plain agar with lamellae, 12–48 (mean: 23), without lamellae 15–49 (mean: 24). Variation as ± 95% confidence limits. Names of compounds in Table 6.4.
[a]Promotion significant at $P = 0.05$. None of the promotions with plain agar (or glucose) differed significantly at $P = 0.05$ from those with the parallel test compounds, except for IAA, which inhibited production of active diffusate.
[b]Receiver and control blocks with 0.05 M glucose.

Table 6.6. Growth Promotion in Decapitated *Flammulina velutipes* Stipes by Lamellar Diffusate Passed Through Dialysis Membranes and by Boiled Diffusate.

Treatments	Elongation (cm)	Promotion over controls without diffusate (cm)
24-h diffusion from lamellae through dialysis membrane into PDA/2	1.91 ± 0.39	0.94[a]
PDA/2 in contact with dialysis membrane for 24 h	0.97 ± 0.11	
24-h lamellar diffusate in PDA/2, boiled 30 min	2.32 ± 0.25	1.31[a]
24-h lamellar diffusate in PDA/2, remelted	2.13 ± 0.26	1.12[a]
PDA/2 boiled 30 min	1.01 ± 0.20	

Initial fruitbody length, 1.5–2.5 cm. Growth period, 3 days. Number of stipes per treatment, 20–27. Variations as ± 95% confidence limits.
[a]Promotion significant at $P = 0.05$.

data from earlier experiments showed that growth promotion (1.04 cm) did not differ significantly from the effects of remelted or boiled diffusate.

Excised lamellae on PDA/2 blocks were placed on glass slides in high humidity at growth room temperature (15–17 °C), and were transferred after different periods to new blocks on test stipes. The lamellae produced considerable active diffusate for at least 28 h but lost most of their activity by 42 h. Strong activity was found in the original receiver blocks with 28-h diffusate but none in 42-h diffusate. Release of active diffusate from the older blocks could have been reduced because of unavoidable shrinkage of the agar through water uptake by the lamellae, but it is unlikely that activity would have been absent for that reason alone. In other experiments diffusates were collected in agar strips from standard amounts of lamellae for 24 and 48 h at 15–17 °C and at 5 °C. The strips were cut into standard blocks for testing under the usual conditions. Growth promotion was the same statistically with 24-h diffusates obtained at both temperatures, but that at 15–17 °C showed a trend towards less promotion. The 24-h diffusates were almost as effective as newly excised lamellae on PDA/2 blocks left on stipes for 3 days. With 48-h diffusate at 5 °C, growth promotion was only about half of that with 24-h diffusate, although there was little shrinkage. The 48-h diffusate at 15–17 °C was even less active. After 48 h on receiver strips at 5 °C, lamellae lost slightly more than 40% of their activity, and at 15–17 °C close to 90% compared to freshly excised lamellae tested at the growth room temperature.

The loss of activity after prolonged diffusion from excised lamellae is being studied further. It cannot be attributed to microbial contamination. Even after several days no colonies were visible on diffusate blocks at either temperature, although some bacteria were observed under high power on preparations from some 15–17 °C blocks. Bacterial colonies appeared in 2 days at 15–17 °C on most PDA/2 blocks left without lamellae. Fruitbody tissues of *F. velutipes* cause only minor pH changes in PDA/2 and apparently have antibacterial action. Antibiotics active against bacteria and *Candida albicans* (Robin) Berkh. have been reported in filtrate from submerged mycelial cultures of *F. velutipes* (Kozová and Řeháček, 1967). Accumulation of an inhibitor or destruction of the growth promoting agent, or both, could occur gradually in receiver blocks during and after prolonged contact with lamellae, and could be more pronounced at higher temperature. Loss of most activity in the blocks and by the lamellae themselves within 2 days at growth room temperature explains in part why lamellar diffusate causes far less than normal stipe elongation.

Other experiments aimed at clarifying the effectiveness of lamellar diffusate included leaving the diffusate in PDA/2 for 1 day on freshly decapitated stipes and replacing it with PDA/2 alone. The growth-promoting effect always persisted after the diffusate had been removed. Similar results were obtained with lamellar diffusate in plain agar. This residual growth promotion is analogous to the residual growth after decapitation of otherwise untreated stipes.

Sensitivity of stipes decapitated at any stage of development decreased as they aged, and active lamellar diffusate did not restore growth of stipes which ceased to elongate.

Discussion

After the early observations of Schmitz (1842) interaction between fruitbody regions in Agaricales received only sporadic attention for almost a hundred years. Recognition of the role of the pileus in stipe elongation was long delayed mainly because the few published decapitation experiments were performed with little or no regard to the stage of development, and without comparative growth measurements on decapitated and intact fruitbodies. Variation in fruitbody size was ignored. Consequently residual growth after decapitation of rapidly elongating stipes was interpreted as indicating continuation of normal growth. Even after Borriss (1934a) first demonstrated the importance of the pileus to stipe growth many years elapsed before the relationship was studied in greater detail. Dependence of stipe elongation on the pileus has now been demonstrated in *Agaricus bisporus, Coprinus radiatus, C. congregatus,* and *Flammulina velutipes.* Recent reports indicate that the general pattern of this relationship is similar in these Agaricales, and that differences are merely a matter of degree or the result of special experimental conditions.

A. bisporus stipes depended on the pileus until approximately the latter half of the phase of rapid elongation when fruitbodies had attained about 70% of their mean final length. *F. velutipes* stipes continued to depend on the pileus until fruitbodies had reached close to 80% of their final length (Table 6.1). These results apply to fruitbodies connected to mycelium. Normal growth of *C. congregatus* stipes attached to mycelium also depended on the pileus almost to the end of the growth period (Bret, 1977a).

In contrast, stipe growth was the same with and without pileus in *Coprinus cinereus* fruitbodies excised just before or during rapid elongation (Gooday, 1974), and excised *C. radiatus* stipes depended on the pileus only until approximately the onset of rapid elongation when they had reached at most one-quarter of their final length (Eilers, 1974). These results were obtained by comparing decapitated and intact excised fruitbodies deprived of exogenous nutrients. Under such conditions starvation itself would reduce elongation of stipes during part of their development. Thus Gooday found that whole *C. cinereus* fruitbodies excised at 10–35 mm grew less than comparable attached ones, and that exogenous nutrients stimulated growth of the excised specimens. In this length range, whole fruitbodies supplied with water alone elongated only as much as excised decapitated stipes. However, elongation may have been the same for the two conditions not because of independence of the stipe from the pileus but because starvation inhibited whole fruitbodies relatively more than decapitated stipes. In *C. congregatus* (Bret, 1977a) whole fruitbodies grew only slightly more than decapitated stipes when both were excised from shortly before until almost the end of the phase of rapid elongation. Excision and lack of exogenous nutrients reduced growth of whole fruitbodies much more than that of decapitated stipes compared, respectively, to attached intact and decapitated specimens. Starvation may reduce production of the lamellar growth regulator. In addition, metabolites available for stipe elongation probably continue to move into the pileus after excision and are lost through sporulation, as CO_2, and in species of *Coprinus* by autolysis. Losses must be smaller in excised decapitated stipes. No reliable conclusions regarding duration of the dependence of normal stipe elongation on the pileus can be drawn from comparisons between starved decapitated and intact fruitbodies.

In young fruitbodies of *Agaricus bisporus* and *Flammulina velutipes* the effects of partial or complete removal of different pileus regions demonstrated that the lamellae are the growth-controlling region. In older fruitbodies of *A. bisporus* comparison between residual growth of decapitated stipes and of stipes with central pilear context left on the apex showed that dependence on the whole pileus (see above) extended over a longer period than dependence on the lamellae. Lamellae were required until at most the middle of the phase of rapid elongation when 55–60% of the mean final length had been reached. A similar limit can be extrapolated from Hagimoto's data (1963, Figs. 1, 2) for stipes with pileus remnants with and without lamellae. No definite explanation can be given for the effect of the context on older stipes. Although diffusate from young peripheral context is inactive, older non-growing context from the middle of the thick *A. bisporus* pileus remains to be tested for possible release of growth-promoting diffusate (see also Role of the Pileus in Stipe Elongation of Agaricales, above). In *F. velutipes* the central pilear context increased residual growth only slightly but significantly for a limited time during rapid elongation. Attached lamellae alone promoted growth even in older fruitbodies (Table 6.1), and as far as can be inferred from surgical experiments the upper limit for dependence of stipe elongation on the pileus coincided with the limit of dependence on the lamellae. In this and other species a certain late stage of development will be reached when stipes contain enough growth-promoting agent from the lamellae to reach their normal final length without depending further on the lamellae.

Expansion of the pilear context is also controlled by the lamellae in *A. bisporus* and probably in *F. velutipes* and species of *Coprinus*. The lamellae themselves expand, but no attempt has been made as yet to locate the region of growth control within these structures.

The special role of the lamellae was confirmed in *A. bisporus* and *F. velutipes* by the fact that they released at least one agar-diffusible substance which increased elongation of decapitated stipes. A diffusible agent was probably active also in *Coprinus radiatus* where stipe elongation was promoted by replaced pilei (Eilers, 1974). The evidence for growth promotion by lamellar diffusate in *A. bisporus* rests solely on curvature tests with diffusate applied mainly to the lateral surface of decapitated stipes or to lamella-free pileus remnants left on the stipe. The activity in diffusate was attributed to several amino acids (Konishi and Hagimoto, 1962; Konishi, 1967), but no supporting data were given, and no further studies have been published on this species.

Work on lamellar diffusate in *F. velutipes* was done with stipes grown and selected by standard procedures, and several test methods were evaluated. Curvature tests proved unsatisfactory for several reasons, including evidence that active diffusate penetrated only slowly from agar through the stipe surface. Application of diffusates on the apex of decapitated stipes demonstrated that a growth-promoting agent was released by the lamellae but not by other fruitbody regions. Lamellar diffusate obtained with or without various nutrients in the receiver blocks always increased elongation of young stipes significantly compared to controls. The activity with potato extract was almost double that with plain agar, and vitamin-free casein hydrolysate (0.02%) had a limited but significant stimulating effect. Growth promotion by lamellae was not increased significantly by glucose alone or by several nitrogen sources with glucose which stimulated growth of excised whole fruitbodies (Gruen and Wu, 1972b). PDA/2 was the outstanding exception because it had a strong effect in both systems. Without lamellar diffusate some nutrient mixtures, including PDA/2, inhibited residual growth of decapitated stipes to some extent, while others had either no effect or, like glucose, caused weak promotion. No nutrient tested replaced the role of the lamellae. However, in testing various nutrients on lamellae, diffusate and nutrients in the receiver blocks were applied together to the stipes. The increased response with potato extract could have resulted from increased production of the growth regulator by the lamellae, or at least partly from the combined effect of lamellar factor and nutrient on the stipe. The former hypothesis is favored because diffusate in plain agar applied to the stipe apex gave the same promotion with plain agar or PDA/2 applied to the lateral surface in rings (Table 6.6). The evidence is not yet conclusive because the speed of nutrient penetration from agar into the stipe surface requires further study. Other indirect evidence against interaction between lamellar diffusate and potato extract in the stipe was the inhibition of residual growth by PDA/2 alone compared to plain agar alone. If there were interactions, at least limited promotion would have been expected with PDA/2 because residual amounts of the lamellar growth factor were probably present in the stipe after decapitation (see Some Characteristics of the Lamellar Diffusate, above).

Too little is known about the nutritional relationship between lamellae and the rest of the fruitbody to explain the effect of potato extract on production of the growth-promoting factor. Young excised lamellae lose most of their capacity to produce the active agent in about two days even in the presence of PDA/2, but production continues in the fruitbody for several days and probably depends on a continuous supply of unknown materials transported from the stipe. Pilear context cannot be the source. Evidence linking the nutritional role of the mycelium in stipe elongation to the controlling role of the lamellae is only circumstantial at present. Nutrients increase growth of excised whole fruitbodies most at the beginning of rapid elongation when they are most dependent on the mycelium. This was indicated by the low ratio of residual growth of excised specimens on water to the normal growth of fruitbodies connected to mycelium (Gruen and Wu, 1972b). At the same stage the stipes are also strongly dependent on the lamellae and the lamellar growth factor is most effective on the stipe. Interestingly, this stage also coincides with the beginning of spore discharge in *F. velutipes* at about 2 cm fruitbody length. With increasing age fruitbody elongation becomes progressively more independent of the mycelium and exogenous nutrients less effective. The work on excised whole fruitbodies was done with specimens grown on PDA, which elongate less than those grown on sawdust medium. Thus, only the general trends in the upper limits for dependence on mycelium and pileus are comparable. However, potato glucose solution ceases to promote elongation of whole fruitbodies close to the end of the phase of rapid elongation, or approximately at the stage (about 70% of final length) when the growth-promoting activity of PDA/2-fed lamellae becomes very low. Probably by that time enough nutrients are present in the fruitbody to supply both the substrates needed for cell wall synthesis and the precursors for synthesis of the growth regulator as long as this continues in the lamellae during the remainder of the growth period.

It is still unknown whether more than one substance is responsible for the growth-promoting effect of the lamellae and whether the same agent functions in different species. The agent is probably active at low concentration because diffusate from small amounts of lamellae in *Agaricus bisporus* and *Flammulina velutipes* (Gruen, 1976) suffices to give a response in decapitated stipes. Heat stability and a molecular weight of less than 12,000 suggest that the active agent in *F. velutipes* is not a protein. The agent is transported from the lamellae into the stipe and pilear context. Movement into the stipe occurs in the hyphal layer at the juncture between pilear context and lamellae. The effects of the lamellar growth factor cannot be replaced by a number of nutrients, by cAMP, or by known plant growth regulators such as IAA, gibberellic acid, or cytokinin. Some of the characteristics of the lamellar agent can be regarded as those of a growth hormone, namely origin in a specific fruitbody region, growth-promoting action on different regions, and effectiveness at low concentration. On the other hand, the material does not induce resumption of growth in decapitated stipes which ceased to elongate. It is preferable at present to designate the active material by a less specific term such as growth-promoting factor or agent, growth regulator, or hormone-like substance.

Acknowledgments. The author wishes to thank Mrs. Eva Wong for her technical assistance in the work on *Flammulina velutipes* and Mr. Alex Campbell for preparing the photographs. The author is grateful to Prof. R. A. A. Morrall for reading the manuscript and for offering helpful comments and criticisms.

The work was supported by Grant A-2371 from the National Research Council of Canada.

References

Almoslechner, E.: Die Hefe als Indikator für Wuchsstoffe. Planta (Berl.) *22*, 515–542 (1934).
Aschan-Åberg, K.: The production of fruit bodies in *Collybia velutipes*. II. Further studies on the influence of different culture conditions. Physiol. Plant. *11*, 312–328 (1958).
Barksdale, A. W.: Sexual hormones of *Achlya* and other fungi. Science *166*, 831–837 (1969).
Bistis, G.: Sexuality in *Ascobolus stercorarius*. I. Morphology of the ascogonium; plasmogamy; evidence for a sexual hormonal mechanism. Am. J. Bot. *43*, 389–394 (1956).

Bistis, G.: Sexuality in *Ascobolus stercorarius*. II. Preliminary experiments on various aspects of the sexual process. Am. J. Bot. *44*, 436–443 (1957).

Borriss, H.: Beiträge zur Wachstums- und Entwicklungsphysiologie der Fruchtkörper von *Coprinus lagopus*. Planta (Berl.) *22*, 28–69 (1934a).

Borriss, H.: Über den Einfluss äusserer Faktoren auf Wachstum und Entwicklung der Fruchtkörper von *Corprinus lagopus*. Planta (Berl.) *22*, 644–684 (1934b).

Bouillenne-Walrand, M., Engels, L., Willam, A.: Teneur en substances de croissance du champignon de couche (*Agaricus hortensis* C. var. *alba*). Mushroom Sci. *2,* 26–28 (1953).

Brefeld, O.: Botanische Untersuchungen über Schimmelpilze, III. Heft: Basidiomyceten I. Leipzig: Arthur Felix 1877a.

Brefeld, O.: Ueber die Bedeutung des Lichtes für die Entwickelung der Pilze. (I. Mittheilung.) Sitz. Ber. Ges. Naturf. Freunde Berlin 1877, 127–136 (1877b).

Bret, J. P.: Respective role of cap and mycelium on stipe elongation of *Coprinus congregatus*. Trans. Br. Mycol. Soc. *68,* 363–369 (1977a).

Bret, J. P.: Release of an inhibiting stipe elongation substance by continuous light grown fruit body caps of *Coprinus congregatus*. Second Int. Mycol. Congress, Tampa, Fla., Abstr. Vol. A-L, p. 67 (1977b).

Buller, A. H. R.: Researches on Fungi, Vol. 1. London: Longmans, Green and Co. 1909.

Bu'Lock, J. D.: Hormones in fungi. In: The Filamentous Fungi. Vol. 2, Biosynthesis and Metabolism. Smith, J. E., Berry, D. R. (eds.). New York: John Wiley & Sons 1976, pp. 345–368.

Butler, G. M.: Growth of hyphal branching systems in *Coprinus disseminatus*. Ann. Bot. (Lond.), N.S. *25,* 341–352 (1961).

Cox, R. J., Niederpruem, D. J.: Differentiation in *Coprinus lagopus* III. Expansion in excised fruit-bodies. Arch. Microbiol. *105,* 257–260 (1975).

Craig, G. D., Gull, K., Wood, D. A.: Stipe elongation in *Agaricus bisporus*. J. Gen. Microbiol. *102,* 337–347 (1977).

Durand, R. J. C.: Interaction of light and temperature on fruitbody morphogenesis in a basidiomycete, *Coprinus congregatus*. Second Int. Mycol. Congress, Tampa, Fla., Abstr. Vol. A–L, p. 151 (1977).

Eilers, F. I.: Growth regulation in *Coprinus radiatus*. Arch. Microbiol. *96,* 353–364 (1974).

Gooday, G. W.: Control of development of excised fruit bodies and stipes of *Coprinus cinereus*. Trans. Br. Mycol. Soc. *62,* 391–399 (1974).

Gräntz, F.: Ueber den Einfluss des Lichtes auf die Entwicklung einiger Pilze. Doctoral Thesis, Unviversität Leipzig, 74 pp. (1898).

Gruen, H. E.: Auxins and fungi. Ann. Rev. Plant Physiol. *10,* 405–440 (1959).

Gruen, H. E.: Endogenous growth regulation in carpophores of *Agaricus bisporus*. Plant Physiol. *38,* 652–666 (1963).

Gruen, H. E.: Growth regulation in fruitbodies of *Agaricus bisporus*. Mushroom Sci. *6,* 103–120 (1967).

Gruen, H. E.: Growth and rotation of *Flammulina velutipes* fruitbodies and the dependence of stipe elongation on the cap. Mycologia *61,* 149–166 (1969).

Gruen, H. E.: Promotion of stipe elongation in *Flammulina velutipes* by a diffusate from excised lamellae supplied with nutrients. Can. J. Bot. *54,* 1306–1315 (1976).

Gruen, H. E.: Control of rapid stipe elongation by the lamellae in fruit bodies of *Flammulina velutipes*. Can. J. Bot. *57,* 1131–1135 (1979).

Gruen, H. E., Wu, S.: Dependence of fruitbody elongation on the mycelium in *Flammulina velutipes*. Mycologia *64,* 995–1007 (1972a).

Gruen, H. E., Wu, S.: Promotion of stipe elongation in isolated *Flammulina velutipes* fruit bodies by carbohydrates, natural extracts, and amino acids. Can. J. Bot. *50*, 803–818 (1972b).
Gyurkó, P.: Die Rolle der Belichtung bei dem Anbau des Austernseitlings *(Pleurotus ostreatus)*. Mushroom Sci. *8*, 461–469 (1972).
Hagimoto, H.: Studies on the growth of fruitbody of fungi IV. The growth of fruit body of *Agaricus bisporus* and the economy of the mushroom growth hormone. Bot. Mag. (Tokyo) *76*, 256–263 (1963).
Hagimoto, H.: On the growth of the fruitbody of *Agaricus bisporus* (Lange) Sing. Trans. Mycol. Soc. Jpn. *4*, 158–164 (1964).
Hagimoto, H., Konishi, M.: Studies on the growth of fruitbody of fungi I. Existence of a hormone active to the growth of fruitbody in *Agaricus bisporus* (Lange) Sing. Bot. Mag. (Tokyo) *72*, 359–366 (1959).
Hagimoto, H., Konishi, M.: Studies on the growth of fruitbody of fungi II. Activity and stability of the growth hormone in the fruitbody of *Agaricus bisporus* (Lange) Sing. Bot. Mag. (Tokyo) *73*, 283–287 (1960).
Jeffreys, D. B., Greulach, V. A.: The nature of tropisms of *Coprinus sterquilinus*. J. Elisha Mitchell Sci. Soc. *72*, 153–158 (1956).
Kamada, T., Miyazaki, S., Takemaru, T.: Quantitative changes of DNA, RNA and protein during basidiocarp maturation in *Coprinus macrorhizus*. Trans. Mycol. Soc. Jpn. *17*, 451–460 (1976).
Kitamoto, Y., Gruen, H. E.: Distribution of cellular carbohydrates during development of the mycelium and fruitbodies of *Flammulina velutipes*. Plant Physiol. *58*, 485–491 (1976).
Knoll, F.: Untersuchungen über Längenwachstum und Geotropismus der Fruchtköperstiele von *Coprinus stiriacus*. Sitzungsber. Kaiserl. Akad. Wiss., Mathem.-Naturwiss. Kl., Abt. I., *118*, 575–634 (1909).
Konishi, M.: Growth promoting effect of certain amino acids on the *Agaricus* fruitbody. Mushroom Sci. *6*, 121–134 (1967).
Konishi, M., Hagimoto, H.: Studies on the growth of fruit body of fungi III. Occurrence, formation and destruction of indole acetic acid in the fruit body of *Agaricus bisporus* (Lange) Sing. Plant Cell Physiol. *2*, 425–434 (1961).
Konishi, M., Hagimoto, H.: Growth-promoting effect of amino acids in the *Agaricus* fruitbody. Plant Physiol. suppl. *37*, ix–x (1962).
Kozová, J., Řeháček, Z.: Antibiotics of *Flammulina velutipes* cultivated in submerged culture. Folia Microbiol. *12*, 567–568 (1967).
Larpent, J. P.: Caractères et déterminisme des corrélations d'inhibition dans le mycélium jeune de quelques champignons. Ann. Sci. Nat. Bot. Biol. Vég. 12ᵉ Sér. *7*, 1–130 (1966).
Magnus, W.: Über die Formbildung der Hutpilze. Arch. Biontol. *1*, 85–161 (1906).
Manachère, G.: Recherches physiologiques sur la fructification de *Coprinus congregatus* Bull. ex Fr.: action de la lumière; rythme de production de carpophores. Ann. Sci. Nat. Bot. Biol. Vég. 12ᵉ Sér *11*, 1–96 (1970).
Matthews, T. R., Niederpruem, D. J.: Differentiation in *Coprinus lagopus* I. Control of fruiting and cytology of initial events. Arch. Mikrobiol. *87*, 257–268 (1972).
Michalenko, G. O.: The assay of growth-promoting substances in *Flammulina velutipes* fruitbodies with a standardized stipe curvature test. Ph.D. Thesis, Univ. of Saskatchewan, 213 pp. (1971).
Naudy-de Serres, M., Latché, J. C., Baldy, P.: Culture de *Collybia velutipes* (Curt.) Quél. sur milieu synthétique liquide. Analyse de quelques constituants du mycélium

à son premier stade de développement à l'obscurité. C. R. Acad. Sci. (Paris) Sér. D *281*, 259–262 (1975).

Nielson, N.: Über das Vorkommen von Wuchsstoff bei *Boletus edulis*. Biochem. Z. *249*, 196–198 (1932).

Pegg, G. F.: Gibberellin-like substances in the sporophores of *Agaricus bisporus* (Lange) Imbach. J. Exp. Bot. *24*, 675–688 (1973).

Pinto-Lopez, J., Almeida, M. G.: *"Coprinus lagopus"* a confusing name as applied to several species. Port. Acta Biol. Ser. B, *11*, 167–204 [1972] (1970/1971).

Plunkett, B. E.: Nutritional and other aspects of fruitbody production in pure cultures of *Collybia velutipes* (Curt.) Fr. Ann. Bot. (Lond.) N.S. *17*, 193–217 (1953).

Reijnders, A. F.: Les Problèmes du Développement des Carpophores des Agaricales et de Quelques Groupes Voisins. Dr. W. Junk, Den Haag 1963.

Robert, J. C.: Fruiting of *Coprinus congregatus:* biochemical changes in fruitbodies during morphogenesis. Trans. Br. Mycol. Soc. *68*, 379–387 (1977).

Rypáček, V., Sladký, Z.: The character of endogenous growth regulators in the course of development in the fungus *Lentinus tigrinus*. Mycopathol. Mycol. Appl. *46*, 65–72 (1972).

Rypáček, V., Sladký Z.: Relation between the level of endogenous growth regulators and the differentiation of the fungus *Lentinus tigrinus* studied in a synthetic medium. Biol. Plant. (Prague) *15*, 20–26 (1973).

Schmitz, J.: Mykologische Beobachtungen, als Beiträge zur Lebens- und Entwicklungsgeschichte einiger Schwämme aus der Klasse der Gastromyceten und Hymenomyceten. Linnaea *16*, 141–215 (1842).

Schwantes, H. O., Hagemann, F.: Untersuchungen zur Fruchtkörperbildung bei *Lentinus tigrinus* Bull. Ber. Dtsch. Bot. Ges. *78*, 1. Generalversammlungsheft, (89)–(101) (1965).

Sladký, Z., Tichý, V.: Stimulation of the formation of fruiting bodies of the fungus *Lentinus tigrinus* (Bull.) Fr. by growth regulators. Biol. Plant. (Prague) *16*, 436–443 (1974).

Streeter, S. G.: The influence of gravity on the direction of growth of *Amanita*. Bot. Gaz. *48*, 414–426 (1909).

Szabó, L. G., Pozsár, B. I., Kota, M.: Cytokinin activity of the fruiting body of *Coprinus micaceus* Fr. Acta Agron. Acad. Sci. Hung. *19*, 402–403 (1970).

Takemaru, T., Kamada, T.: Basidiocarp development in *Coprinus macrorhizus* I. Induction of developmental variations. Bot. Mag. (Tokyo) *85*, 51–57 (1972).

Turner, E. M.: Development of excised sporocarps of *Agaricus bisporus* and its control by CO_2. Trans. Br. Mycol. Soc. *69*, 183–186 (1977).

Uno, I., Ishikawa, T.: Chemical and genetical control of induction of monokaryotic fruiting bodies in *Coprinus macrorhizus*. Mol. Gen. Genet. *113*, 228–239 (1971).

Uno, I., Ishikawa, T.: Purification and identification of the fruiting-inducing substances in *Coprinus*. J. Bacteriol. *113*, 1240–1248 (1973a).

Uno, I., Ishikawa, T.: Metabolism of adenosine 3′,5′-cyclic monophosphate and induction of fruiting bodies in *Coprinus macrorhizus*. J. Bacteriol. *113*, 1249–1255 (1973b).

Urayama, T.: Das Wuchshormon des Fruchtkörpers von *Agaricus campestris* L. (Vorläufige Mitteilung). Bot. Mag. (Tokyo) *69*, 298–299 (1956).

van den Ende, H.: Sexual Interactions in Plants. London: Academic Press 1976.

Wong, W. M.: Distribution of endogenous amino acids and proteins during fruitbody development in *Flammulina velutipes* (Curt. ex Fr.) Sing. Ph.D. Thesis, Univ. of Saskatchewan, 169 pp. (1978).

Wong, W. M., Gruen, H. E.: Changes in cell size and nuclear number during elongation of *Flammulina velutipes* fruitbodies. Mycologia *69*, 899–913 (1977).

Chapter 7

Metabolic Control of Fruitbody Morphogenesis in *Coprinus cinereus*

GRAHAM W. GOODAY

Introduction

A study of the morphogenesis of the agaric fruitbody, as well as being of fundamental mycological interest, has relevance to commercial mushroom production, where it could provide information to aid increases in synchrony and productivity of fruiting. It also provides a model system for the elucidation of the controls of differentiation in multicellular eukaryotes.

The account that follows is of some aspects of the elongation of the stipe of *Coprinus cinereus* (Schaeff. ex Fr.) S. F. Gray. This can occur in excised stipes in the absence of exogenous nutrients or water (Gooday, 1974). The work has been directed towards determining which of the observed changes in levels of chemical components of the cells are essential for elongation, and from this to determining which are limiting, and thus controlling, the differentiation process.

Moore et al. (1979) provide an excellent comprehensive review of many aspects of the development of fruitbodies of *Coprinus cinereus*. Their review includes an assessment of the nomenclature of *C. cinereus*, which they conclude was the species used by some Japanese workers and ascribed by these workers as *Coprinus macrorhizus* Rea f. *microsporus* Hongo.

Materials and Methods

The dikaryon of *Coprinus cinereus*, C692 × MAE131, was grown as described by Gooday (1974). Stipes were harvested, freed from cap and basal mycelium, measured to the nearest millimeter, weighed to the nearest 0.1 mg, immediately freeze-dried, and re-weighed. Each stipe was ground to a powder, and extracted thrice with 2 ml methanol:water (4:1, v/v) for at least 15 h at 45 °C. Preliminary checks had confirmed that this procedure removed all detectable glucose and trehalose. The extracts were combined, dried under vacuum, and dissolved in 1 ml water for analysis for glucose and trehalose. Glucose was estimated on a sample by the glucose oxidase method using ABTS as redox indicator (Boehringer Corp., London). A further sample (0.1 ml) was treated with a solution of freeze-dried snail gut juice ("helicase"; L'Industrie Biologique Francaise, Paris; 5 mg ml^{-1}; in

0.1 M potassium phosphate buffer, pH 7.0) for 18 h at 30 °C, with a drop of chloroform to maintain sterility. Preliminary experiments had established that these were optimum conditions for the hydrolysis of trehalose by the trehalase in the snail gut juice. The glucose content of the digest was then measured as before, and the trehalose content of the extract estimated by subtracting the value obtained for the free glucose, and correcting the trehalose content by reference to the hydrolysis of standard solutions of trehalose by the snail gut juice. Previous experiments had confirmed the absence of other oligosaccharides that might also give rise to glucose on incubation with this enzyme mixture, such as glycogen, maltose, and sucrose.

Some stipe residues were then analyzed for glycogen, others for protein.

The glycogen was estimated by treating the methanol-extracted residue with 0.25 ml water at 100 °C for 10 min, and digesting the mixture three successive times with 0.25 ml of a solution of amyloglucosidase (from *Rhizopus* sp.; Sigma Chemical Co., London; 5 mg ml^{-1}; in 0.03 M sodium citrate–phosphate buffer, pH 4.5) for 2 h at 45 °C. Each time, the suspension was centrifuged, and the supernatants were combined. The glucose produced was estimated as above, and the glycogen content calculated using a standard of pure oyster glycogen (Hopkins & Williams Ltd., England). The amyloglucosidase had no lytic activity on other glucans, such as laminarin, sclerotan, S-glucan, or cellulose.

Protein contents in other methanol-extracted stipe residues were measured on alkali digests as described by Lowry et al. (1951), using bovine serum albumen as a standard.

The residues remaining after glycogen or protein estimations were washed with acetone, dried, and soaked overnight at 20 °C in 0.75 ml 6 N HCl in sealed tubes ("Reacti-vial", Pierce Chemical Co., Illinois). The contents of the tubes were well mixed, and heated for 8 h at 100 °C, with hourly re-mixing. The tubes were centrifuged and the glucosamine contents of the supernatants were measured (Tracey, 1955). The chitin contents of the stipes were calculated using controls of hydrolysed purified chitin as standards.

Excised stipes were collected and allowed to elongate as described by Gooday (1974).

Results

Composition of Intact Stipes

All stipes used for this work were of the first crop, growing under uniform conditions. In an attempt to define those processes most closely concerned with elongation there was no selection, i.e., all stipes were used. At least beyond 20 mm long, there was a wide variation in dry weight per unit length (Fig. 7.1). However, it is clear that in general there was an increase in biomass during elongation *in vivo,* presumably chiefly resulting from translocation of nutrients from the vege-

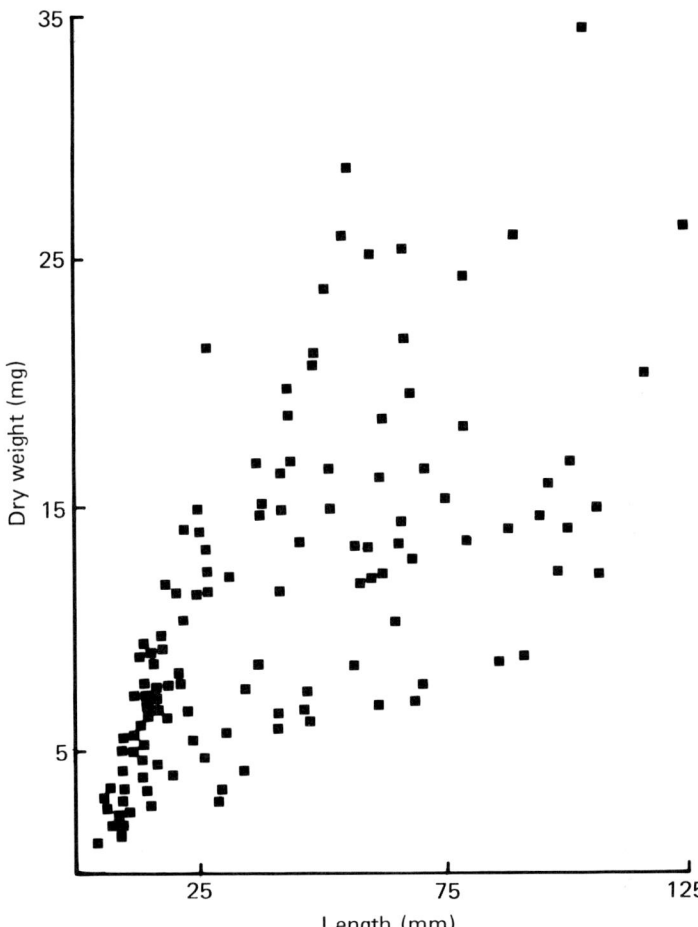

Fig. 7.1. Measurements of dry weight and length for stipes of *Coprinus cinereus* used in this investigation.

tative mycelium. Likewise, increasing curvature of the plot of fresh weight against dry weight (Fig. 7.2) suggested an influx of water from the vegetative mycelium.

Trehalose was a major component of the stipes, and its content increased so that it remained accounting for at least 10% of the dry weight (Fig. 7.3). Glucose contents were always much less, at about 1% at the most, and showed a wide variation. Pooling results for stipes of different lengths, and calculating minimum concentrations from the water contents of the stipes, suggested that the trehalose concentration steadily fell during elongation (Fig. 7.4). The equivalent apparent glucose concentration fell more rapidly to a very much lower figure, but showed a slight rise in the longest stipes, perhaps due to some autolysis.

Glycogen contents of stipes showed very wide variation (Fig. 7.5), but were very low in the larger stipes.

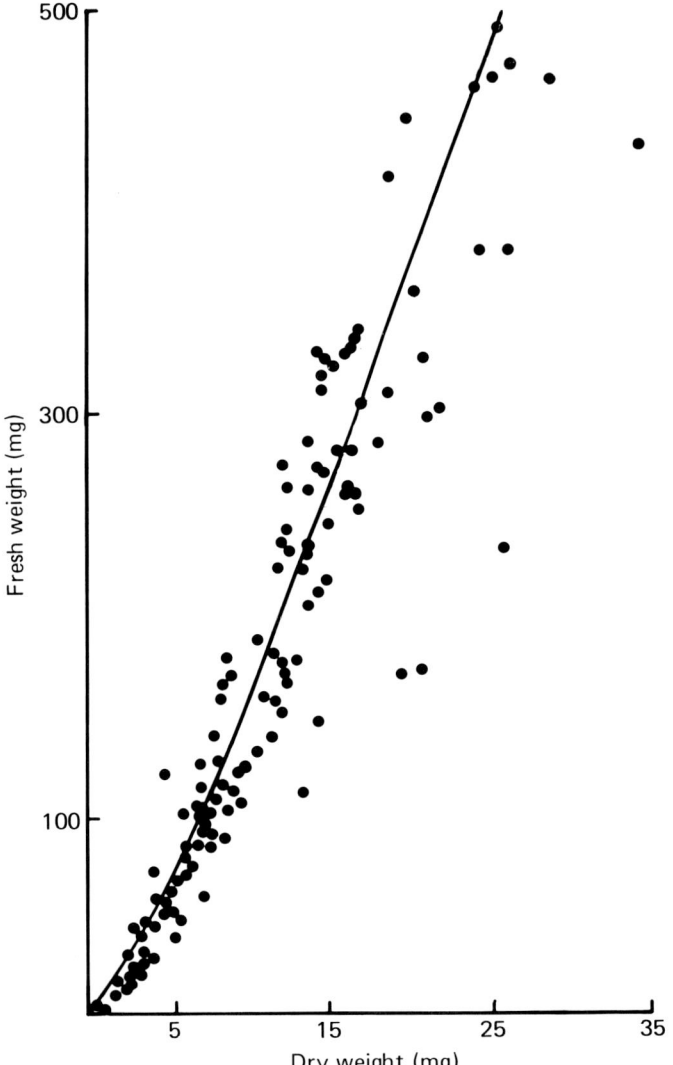

Fig. 7.2. Measurements of fresh weight and dry weight for stipes.

Protein contents showed a steady increase in increasing dry weight of stipe, to remain at about a constant 13.5% of the total (Fig. 7.6).

The chitin content also showed a steady increase, remaining at about 10.9% of the total (Fig. 7.7). This was equivalent to an increment of about 26 μg/mm increase in length.

Addition of the cell components measured here for stipes of different length classes showed that they accounted for nearly 70% of the shortest stipes, and nearly 50% of the longest stipes (Fig. 7.8).

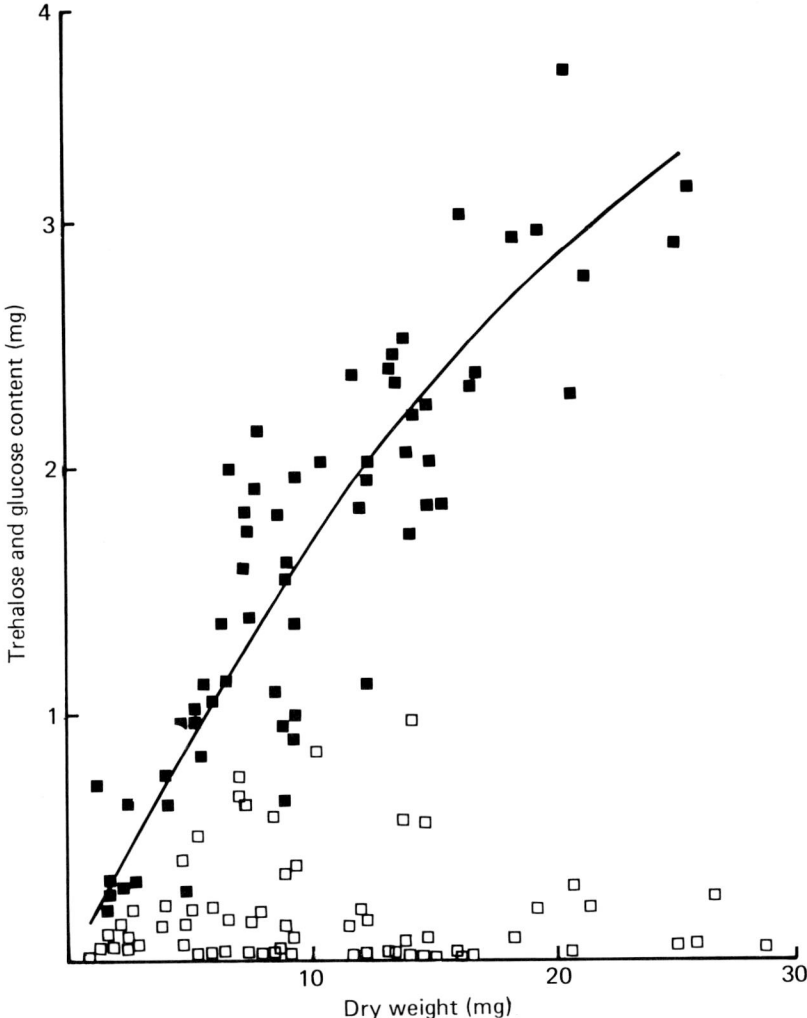

Fig. 7.3. Contents of trehalose (■) and glucose (□) in stipes.

Composition of Elongating Excised Stipes

Excised young stipes of *C. cinereus* elongate when incubated at 25 °C, without the uptake of water or nutrients (Gooday, 1974). Thus, clearly the process of elongation *per se* has no absolute requirement for exogenous materials, or for the net increase in fresh weight, dry weight, and water content that are observed *in vivo*. The smaller stipes elongate less than the larger ones, presumably because they have fewer reserves on which to draw for their endotrophic elongation.

Table 7.1 compares 6 stipes allowed to elongate *in vitro* with 6 control stipes, freshly excised. The 3 smaller elongated stipes, average dry weight 6.0 mg,

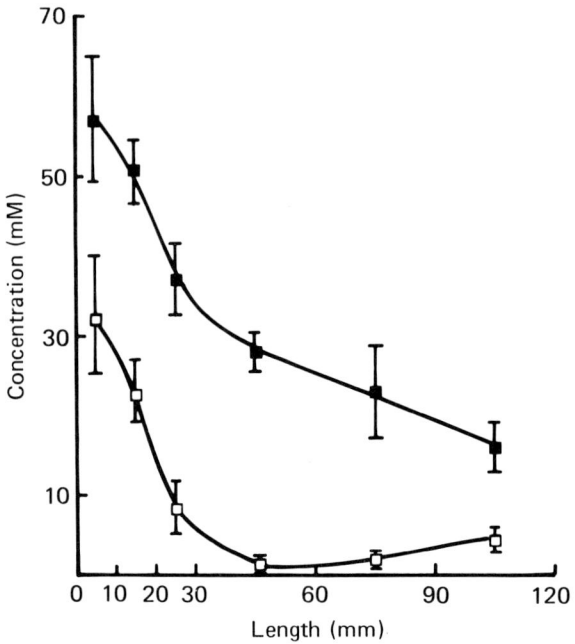

Fig. 7.4. Calculated minimum values for concentrations of trehalose (■) and glucose (□) in stipes.

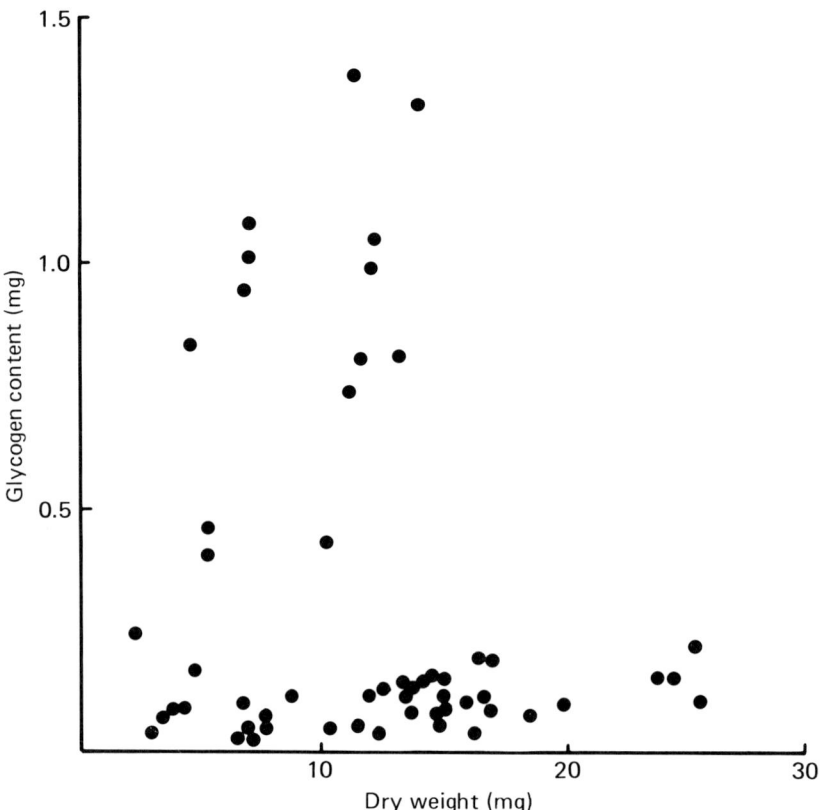

Fig. 7.5. Contents of glycogen in stipes.

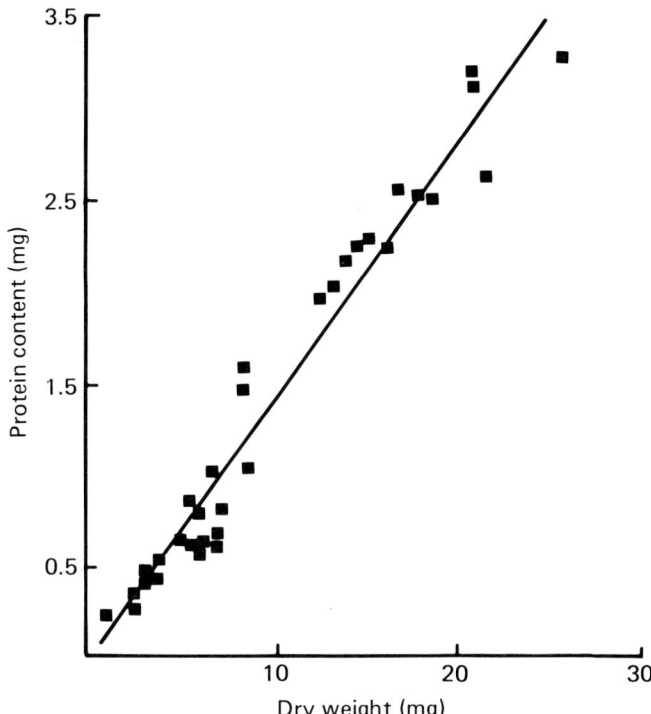

Fig. 7.6. Contents of protein in stipes.

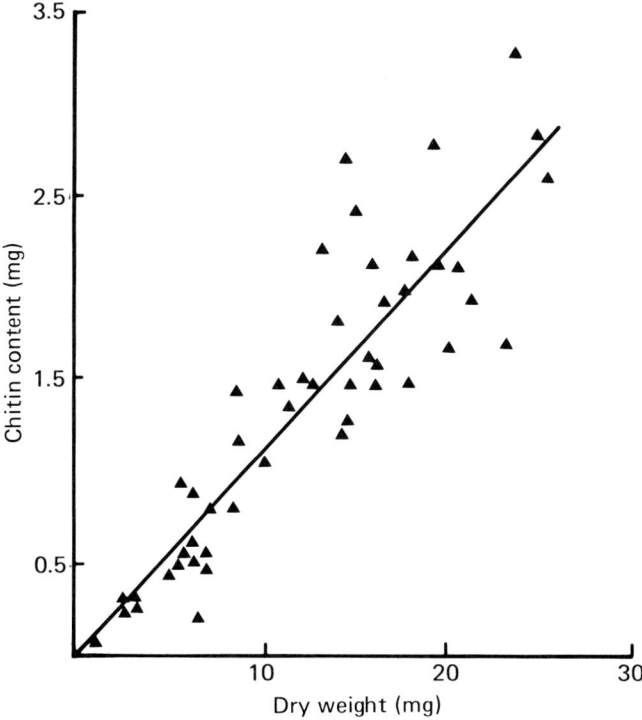

Fig. 7.7. Contents of chitin in stipes.

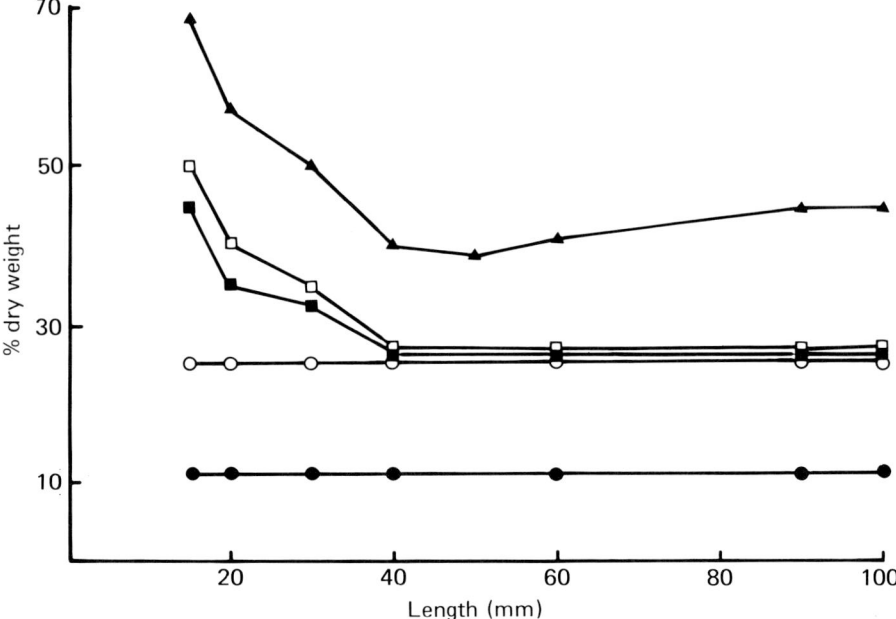

Fig. 7.8. Accumulative totals of percentage dry weights of stipes accounted for by chitin (●), protein (○), glycogen (■), glucose (□), and trehalose (▲).

showed an average increase of 26 mm, whereas the 3 larger ones, average 11.2 mg, elongated by an average of 46 mm. The major component that was observed to increase to keep approximately in step with the elongation was chitin. Thus the control stipes in Table 7.1 had an average chitin content of 10.4 ± 1.8% dry weight (cf. Fig. 7.7), whereas the elongated stipes had an average content of 16.1 ± 1.4% dry weight. The 3 larger stipes had greater increases in chitin than the 3 smaller ones. The increases were equivalent to an increment of about 20 μg/mm increase in length.

Synthesis of Chitin in Stipes

Synthesis of chitin in elongating stipes was investigated by incubating them with radioactive N-acetylglucosamine. This sugar is efficiently incorporated into chitin, presumably by being phosphorylated by N-acetylglucosamine kinase to N-acetylglucosamine-6-phosphate, which can be converted to UDP-N-acetylglucosamine. This nucleotide sugar is then the specific precursor of chitin, and also of some glycoproteins which, however, are relatively minor components of fungal cells.

After injecting two elongating stipes with N-acetyl-[1-^{14}C]-D-glucosamine, over 80% of the radioactivity was associated with the chitin-rich residue left after exhaustive extractions (Table 7.2). Acid hydrolysis in 6 N HCl at 100 °C for 8 h gave material with chromatographic properties of glucosamine as the sole prod-

Table 7.1. Chitin Contents of Excised Stipes of *Coprinus cinereus*.

	Stipes allowed to elongate in vitro[a]				Control stipes, freshly excised		
Length before (mm)	Length after (mm)	Dry weight (mg)	Chitin (mg)		Length (mm)	Dry weight (mg)	Chitin (mg)
15	50	6.9	1.14		15	5.7	0.44
16	33	4.6	0.76		17	4.9	0.56
20	47	6.6	0.98		20	6.0	0.70
33	75	9.4	1.64		32	7.3	0.76
35	75	10.5	1.80		35	8.9	0.76
36	91	13.6	1.90		35	10.7	1.32

[a]Stipes excised, measured, and allowed to elongate for 18 h at 25 °C with their bases dipping in water.

Table 7.2. Incorporation of N-acetylglucosamine into Stipe Tissue of *Coprinus cinereus*. Two stipes were excised and incubated, A in moist air, B with base dipping in water. Each was injected with 25 μl N-acetyl[1^{14}C]-D-glucosamine (250 nCi, 4.4 nmol) into its hollow center and allowed to elongate (15 h, 25 °C, each went from 31 to 82 mm). Stipes were then harvested, homogenized, and extracted successively. Each extract was dried, neutralized where appropriate, and counted in scintillant (10 ml methanol:toluene 2:3, v/v) with butyl-PBD (0.7%, w/v) and Cab-o-sil (2.5%, w/v)).

Fraction	% Total d.p.m. recovered		Probable major radioactive components
	Stipe A	Stipe B	
Ethanol:H$_2$O (6:4, v/v), 50 °C, 1h	7.2	6.0	GlcNAc, UDP-GlcNAc
H$_2$O, 100 °C, 1 h	1.0	0.9	Glycoprotein
KOH:H$_2$O (5:95, w/v), 20 °C, 18 h	1.6	1.7	Glycoprotein
Acetic acid:H$_2$O (5:95, v/v), 20 °C, 1 h	0	0	—
30 vols. H$_2$O$_2$:acetic acid (1:1, v/v), 70 °C, 2 h	8.8	7.1	Degraded chitin
Residue, ultrasonically dispered in ethanol	81.4	84.3	Chitin

uct. The soluble pool extracts were chromatographed [t.l.c., silica gel F254, n-propanol:water 7:3 (v/v)] and about 40% of the label co-chromatographed with free N-acetylglucosamine. Perhaps this was unincorporated precursor from the hollow center of the stipe; perhaps it represented the action of autolytic enzymes in the tissue. About 10% co-chromotographed with UDP-N-acetylglucosamine; the remainder was not identified. The label solubilized by acidic hydrogen peroxide probably represented some degraded chitin (cf. Tracey, 1955).

Light-microscopic autoradiographs of elongating stipe tissue briefly incubated with tritiated N-acetylglucosamine showed that the stipe cells had incorporated radioactivity into their walls (Fig. 7.9). Most of this labeling representing chitin deposition was uniform along the walls, in marked contrast to the apical labeling of chitin deposition in growing hyphae (Gooday, 1971). A few narrower hyphae did show such apical labeling, but probably represented secondary hyphae growing through the stipe tissue.

Electron-microscopic autoradiography confirmed that elongating stipes incorporated exogenously supplied N-acetylglucosamine into their cell walls (Table 7.3, Figs. 7.11–7.12). The specificity of incorporation of this substrate suggests that this was chiefly as chitin. Counts of the distribution of silver grains showed 83.3% to be associated with the walls/cell membranes of the cells. Preparation of the material involved fixation and washing which would remove soluble precursors. As the incubation was short (10 min), the results suggest that there is little or no prepolymerization of chitin at sites remote from the walls. Some silver grains were associated with lomasomes (Fig. 7.12), but analysis showed that their relative intensity of labeling was little higher than that of the cytoplasm (Table 7.3). Lomasomes often appear in areas associated with wall synthesis in fungi, but their significance is unclear at present (Beckett et al., 1974). Like bacterial mesosomes, they may be a physical consequence of the state of the cell membrane at particular sites rather than being functional structures in their own right.

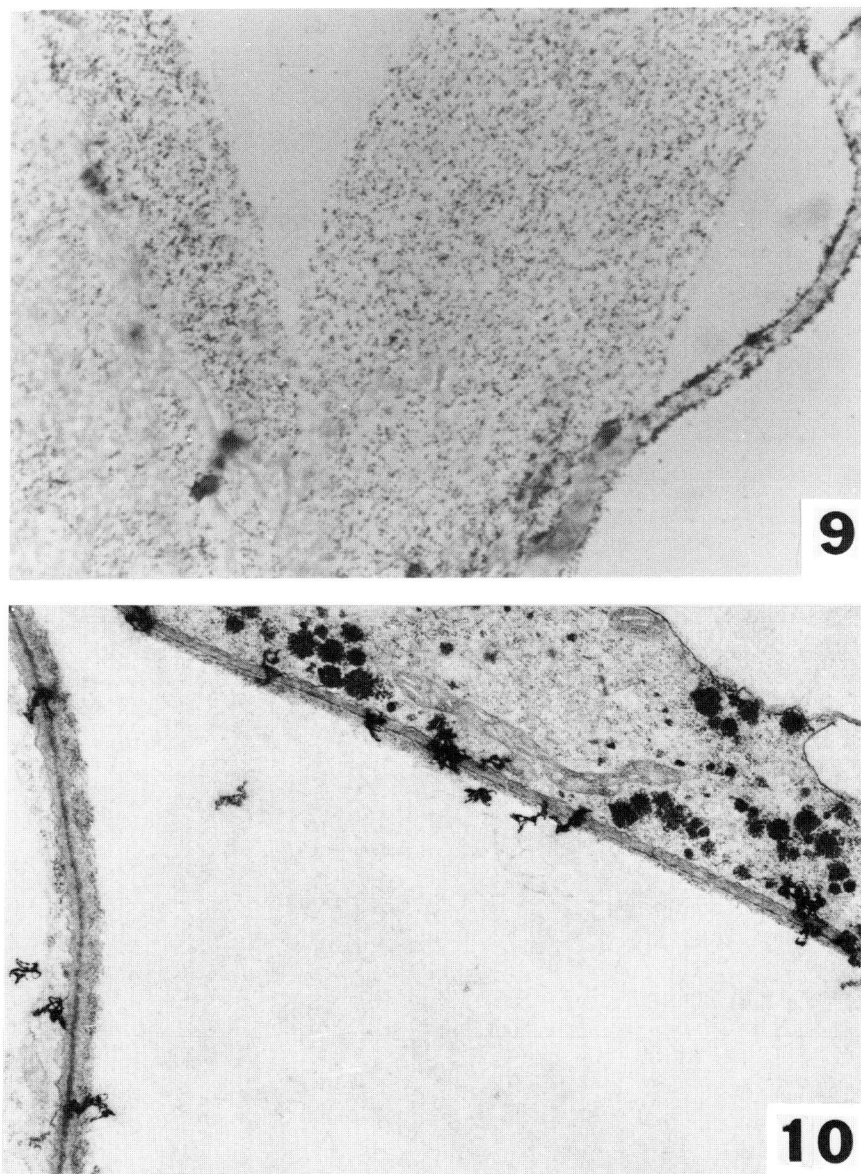

Fig. 7.9. Light-microscopic autoradiograph of incorporation of N-acetyl-[^{3}H]-glucosamine by elongating stipe tissue of *Coprinus cinereus*. Incubated for 10 min, 25°C, teased out and extracted with ethanolic KOH to remove cytoplasm; autoradiography as described by Hunsley and Gooday (1974). Note uniform distribution of silver grains along the walls. Micrograph no. G7.12, × 600.

Fig. 7.10. Electron-microscopic autoradiograph of incorporation of N-acetyl-[^{3}H]-glucosamine by elongating stipe tissue of *Coprinus cinereus*. Incubation as for Fig. 7.9, but for 20 min; autoradiography as described by Hunsley and Gooday (1974). Note silver grains associated with walls of the two cells. Micrograph no. OR.98, × 14,175.

Table 7.3. Electron Microscopic Autoradiography of Chitin formation in Elongating Stipes of *Coprinus cinereus*. Results from segments of elongating stipe tissue (1 mm thick, 20–25 mm from base of 30-mm stipe) incubated for 10 min in N-acetyl-[1-^{3}H]-D-glucosamine. Expressed as silver grain distributions compared to "random circle" distributions on micrographs ($\times$ 22,500). Treatment of material, preparation of micrographs and criteria of assessment were as described by Hunsley and Gooday (1974).

Location	Grain counts		"Random circles"		Relative intensity of labelling[a]
	Actual	%	Actual	%	
Wall	1749	83.3	258	17.5	65.8
Cytoplasm	204	9.7	392	26.6	5.0
Nuclei, mitochondria	9	0.4	48	3.2	1.7
Lomasomes	26	1.2	32	2.2	7.5
Intercellular space	54	2.6	174	11.8	3.0
Vacuoles	58	2.8	571	38.7	1.0
Totals	2100	100.0	1475	100.0	

[a]Calculated as $\dfrac{\text{grain count \%}}{\text{random circle \%}} \times \dfrac{\text{vacuole random circle \%}}{\text{vacuole grain count \%}}$ χ^2 comparison of distribution = 305.2, i.e., the distributions are unrelated and the silver grains have a non-random distribution.

Discussion

General Observations on Elongation

Stipe elongation of *Coprinus cinereus* is accompanied *in vivo* by increases in dry weight and water content. This agrees with previous results for *Coprinus cinereus* (Madelin, 1956; Blayney and Marchant, 1977) and *Coprinus congregatus* Bull. ex Fr. (Bret, 1977; Robert, 1977a, 1977b), where these increases have been observed to be supported by transport of nutrients and water from the basal vegetative mycelium into the stipe.

However, as at least some elongation (quantitatively dependent on species and state of growth) can occur in excised stipes of *C. cinereus* (Gooday, 1974; Cox and Niederpruem, 1975), *Coprinus radiatus* (Bolt. ex Fr.) S. F. Gray (Eilers, 1974), and *C. congregatus* (Bret, 1977), such an influx is not *essential* for elongation. Elongation is a result of divisions as well as elongations of individual cells in *C. radiatus* (Eilers, 1974), *Agaricus bisporus* (Lange) Imbach (Craig et al., 1977) and *Flammulina velutipes* (Curt. ex Fr.) Sing. (Wong and Gruen, 1977). In the case of excised stipes of *C. cinereus*, which can show very considerable elongation in the absence of added water or nutrients, the increases in materials and water required by individual cells for their division and elongation must be supplied by the autolysis and collapse of neighboring cells. This has not been investigated quantitatively, but the stipes are quite hollow after elongation.

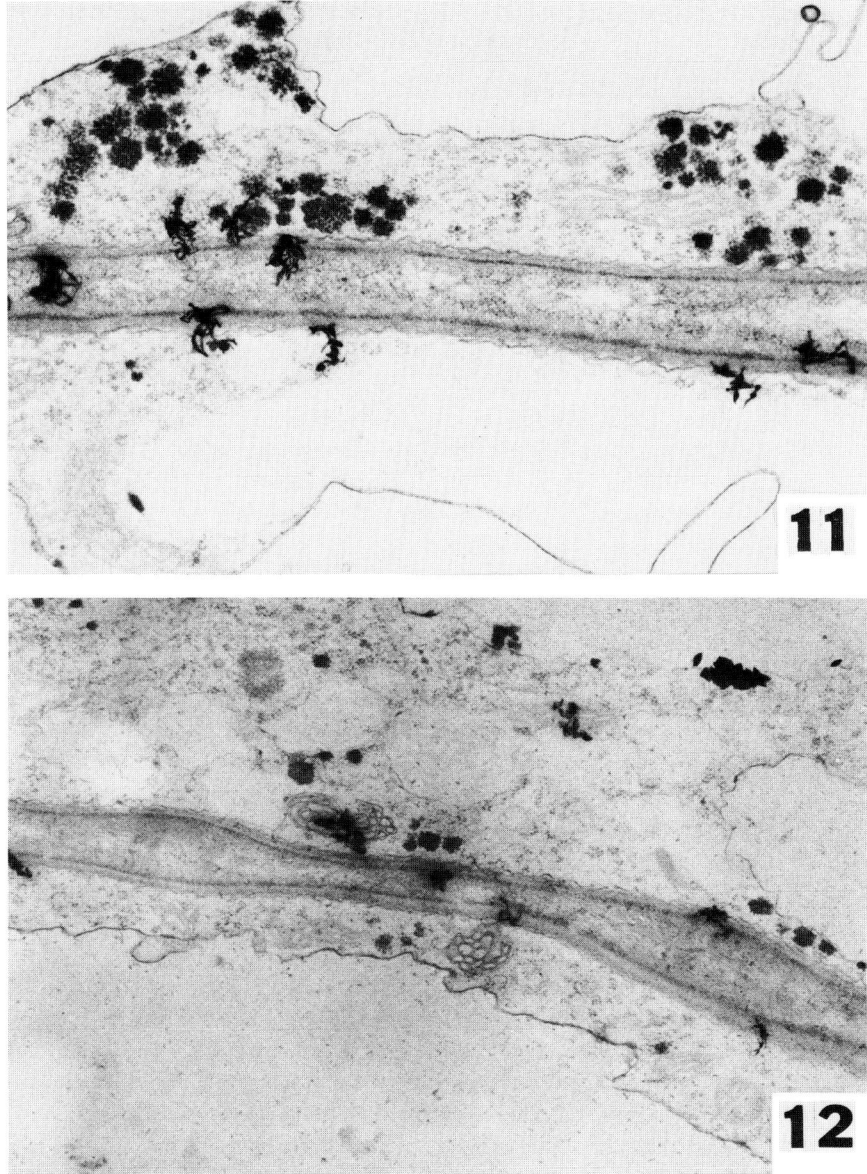

Fig. 7.11. As for Fig. 7.10. Note silver grains associated with walls of the two cells. Micrograph no. OR.99, × 22,500.

Fig. 7.12. As for Fig. 7.10, but incubation for 10 min. Note silver grains associated with lomasome. Micrograph no. OR.70, × 15,000.

Trehalose and Glucose

Trehalose is a common storage carbohydrate in fungi. Two properties that might favor its use as such are that it is non-reducing, and that it requires no energy for its hydrolysis to glucose. It is hydrolysed to glucose by the enzyme trehalase, two forms of which have been described from *Coprinus cinereus* (= *C. lagopus* sensu Buller) (Rao and Niederpruem, 1969). Rao and Niederpruem also present analyses of trehalose and glucose in stipes during elongation, and the results presented here are in general agreement with these, with the trehalose content rising and the glucose content falling during elongation. Changes in levels of these two carbohydrates have been reported for stipes of *Flammulina velutipes,* where trehalose, a major component, rises and then falls during elongation, while glucose, a minor component, shows a slight rise (Kitamoto and Gruen, 1976) and of *Agaricus bisporus,* where trehalose falls from nearly 5% dry weight to about 1%, while glucose rises from a negligible amount to nearly 1% dry weight (Hammond and Nichols, 1976). In *A. bisporus* it is suggested that the drop in trehalose in the stipe represents translocation to the cap for spore formation.

Glycogen

Glycogen is a common reserve polysaccharide in fungi. In electron micrographs of the *Coprinus cinereus* stipe it appears as darkly staining rosettes (visible in Figs. 7.10–7.12; Matthews and Niederpruem, 1973; McLaughlin, 1974; Blayney and Marchant, 1977). Characteristically it is broken down to provide building blocks for the synthesis of new cell components during differentiation. Thus, in species of *Coprinus,* rich deposits of glycogen in swollen vegetative cells disappear as fruitbody primordia are formed (Madelin, 1960). Blayney and Marchant (1977) illustrate and measure its disappearance from elongating cells in the stipes of *C. cinereus,* and suggest that it is being used as a source of precursors for elongation. The results presented here agree with this conclusion, in that higher glycogen contents were only found in smaller stipes. However, the great variability in glycogen contents, with the occurrence of many small stipes with low glycogen contents, suggests that the initial accumulation of glycogen is not tightly controlled. Thus an initial high content may just reflect the success of that particular primordium in attracting nutrients from the vegetative mycelium rather than being a prerequisite for continued differentiation.

Jirjis and Moore (1976) have characterized the glycogen of *Coprinus cinereus* as being very similar to that from other sources. They demonstrated its accumulation in vegetative mycelium, and its subsequent disappearance concomitant with sclerotium production. However, it also disappeared in conditions not allowing sclerotium formation, and these authors conclude that it should be viewed as a transient carbohydrate store. In the gills of maturing fruitbodies of species of *Coprinus,* dense glycogen deposits in the subhymenium disappear during basidium differentiation, but re-accumulate in the developing spores (Bonner et al., 1957; McLaughlin, 1974).

Protein

The protein contents of the elongating stipes were remarkably constant (Fig. 7.6). Blayney and Marchant (1977) describe large protein crystals in the cytoplasm of stipes of *C. cinereus,* and illustrate how these appear to be steadily digested during elongation. Their analyses of protein contents of centrifuged pellets (30,000 × g, 1 h) of stipe homogenates showed good agreement with these electron microscopic observations. They suggest that these protein crystals are used as precursors for stipe components. From results presented here, as the percentage protein content remains so constant, they would appear to be used to make further proteins, and are also probably supplemented by translocation of protein precursors from the vegetative mycelium. Robert (1977a) reports a fall in protein content in the stipes of his synchronous and uniform fruitbodies of *C. congregatus,* from 11.3% to 8.9% dry weight in the rapid elongation in the 12 h preceding sporulation and autolysis. Kamada et al. (1976) present results of total protein contents of stipes of *C. macrorhizus* f. *microsporus,* for a wild type strain, for an "elongationless" mutant, and for a "sporeless" mutant, as increments between 3 and 18 h of development of their system. Thus, the increments were about 1.5-fold for the first two cultures, compared with 3.7-fold for dry weight, whereas the sporeless mutant, which elongated more than the control, showed a much higher increment of protein (nearly 4-fold) but a higher increment of dry weight (5.6-fold).

Chitin

Chitin is found in the cell walls of basidiomycetes, where it forms a structural layer of microfibrils adjacent to the cell membrane (Hunsley and Burnett, 1970; Michalenko et al., 1976; van der Valk et al., 1977). Its structural role in species of *Coprinus* is indicated by the formation of protoplasts by the sole action of chitinase (Moore, 1975). Marchant (1978) presents analyses of walls of monokaryons and dikaryons of *Coprinus cinereus,* giving a value of 43.3% chitin for a whole wall fraction from stipes. The linear increase in chitin with increasing stipe length, and the observation that elongating stipe cells are actively synthesizing chitin along their length (as shown by measurements of incorporation of radioactive *N*-acetylglucosamine by scintillation counting, light-microscopic and electron-microscopic autoradiography), together suggest that the elongation of individual cells may be in part a consequence of chitin formation. The dimensions of a cell at any point in time are defined by its chitinous shell. To elongate, it must produce more chitin. (There is no evidence in fungal cells for a multinet growth of cell surfaces mediated by the sliding of layers of microfibrils over each other, as occurs with the cellulose component of the walls of higher plants.) The importance of chitin synthesis to elongation is shown by the observation of increases in chitin contents of excised stipes elongating endotrophically, where again there was a linear increase in chitin content, albeit less than that shown *in vivo* (20 µg/mm as opposed to 26 µg/mm). Kamada and Takemaru (1977) also present results showing an increase in chitin content of stipes in their wild type strain and their "elongationless" mutant between 6 and 21 h of development in their system.

Concluding Remarks

The cell components measured here: chitin, protein, glycogen, glucose, and trehalose, accounted for nearly 70% of the total cell dry weight of stipes at the onset of rapid elongation at 15 mm long, but only about 40% of elongating cells, at 50 mm long. Likewise, Robert (1977a), measuring total anthrone-positive carbohydrate and protein during development of *C. congregatus,* found that these together accounted for about 60% of the stipe dry weight at the onset of rapid elongation, but only 46% at its termination, and concludes "that substances not being measured are being synthesized during the elongation phase."

The stipe of an agaric fruit body has only one purpose—to bear the spore-producing cap in a position suitable for the release of the spores. Thus, in species of *Coprinus,* it is positively phototropic and negatively geotropic. In nature, it is of no future use to its producing mycelium—it will disappear rapidly by autolysing and by being digested by other microbes. It is an end in itself, and its components such as the 15% dry weight trehalose cannot be thought of as "reserve materials." In this particular case, perhaps the trehalose serves a role in maintaining the turgor pressure of the cells. Although the cell wall, and in particular the chitin, is the shape-determining element of the stipe cells, the stipe is held erect by the turgor pressure of its cells, as is shown by its collapse after freezing and thawing.

Acknowledgments. I thank my colleagues, D. Hunsley and H. Al-Laebi, for their contributions to the work described here.

References

Beckett, A., Heath, I. B., McLaughlin, D. J.: An Atlas of Fungal Ultrastructure. London: Longman 1974.
Blayney, G. P., Marchant, R.: Glycogen and protein inclusions in elongating stipes of *Coprinus cinereus.* J. Gen. Microbiol. *98,* 467–476 (1977).
Bonner, J. T., Hoffman, A. A., Morioka, W. T., Chiquoine, A. D.: The distribution of polysaccharides and basophilic substances during the development of the mushroom *Coprinus.* Biol. Bull. (Woods Hole) *112,* 1–6 (1957).
Bret, J. P.: Respective role of cap and mycelium on stipe elongation of *Coprinus congregatus.* Trans. Br. Mycol. Soc. *68,* 363–369 (1977).
Cox, R. J., Niederpruem, D. J.: Differentiation in *Coprinus lagopus.* III. Expansion of excised fruit bodies. Arch. Microbiol. *105,* 257–260 (1975).
Craig, G. D., Gull, K., Wood, D. A.: Stipe elongation in *Agaricus bisporus.* J. Gen. Microbiol. *102,* 337–347 (1977).
Eilers, F.: Growth regulation of *Coprinus radiatus.* Arch. Microbiol. *96,* 353–364 (1974).
Gooday, G. W.: An autoradiographic study of hyphal growth in some fungi. J. Gen. Microbiol. *67,* 125–133 (1971).
Gooday, G. W.: Control of development of excised fruit bodies and stipes of *Coprinus cinereus.* Trans. Br. Mycol. Soc. *62,* 391–399 (1974).

Hammond, J. B. W., Nichols, R.: Carbohydrate metabolism in *Agaricus bisporus* (Lange) Sing.: changes in soluble carbohydrates during growth of mycelium and sporophores. J. Gen. Microbiol. *93,* 309–320 (1976).
Hunsley, D., Burnett, J. H.: The ultrastructural architecture of the walls of some hyphal fungi. J. Gen. Microbiol. *62,* 203–218 (1970).
Hunsley, D., Gooday, G. W.: The structure and development of septa in *Neurospora crassa.* Protoplasma *82,* 125–146 (1974).
Jirjis, R. I., Moore, D.: Involvement of glycogen in morphogenesis of *Coprinus cinereus.* J. Gen. Microbiol. *95,* 348–352 (1976).
Kamada, T., Miyazaki, S., Takemaru, T.: Quantitative changes of DNA, RNA, and protein during basidiocarp maturation in *Coprinus macrorhizus.* Trans. Mycol. Soc. Japan *17,* 451–460 (1976).
Kamada, T., Takemaru, T.: Stipe elongation during basidiocarp maturation in *Coprinus macrorhizus:* Changes in polysaccharide composition of stipe cell wall during elongation. Plant Cell Physiol. *18,* 1291–1300 (1977).
Kitamoto, Y., Gruen, H. E.: Distribution of cellular carbohydrates during development of the mycelium and fruit bodies of *Flammulina velutipes.* Plant Physiol. *58,* 485–491 (1976).
Lowry, D. H., Rosebrough, N. J., Farr, A. L., Randall, R. J.: Protein measurement with the Folin phenol reagent. J. Biol. Chem. *193,* 265–275 (1951).
Madelin, M. F.: Studies on the nutrition of *Coprinus lagopus* Fr., especially as affecting fruiting. Ann. Bot. (Lond.) *20,* 307–330 (1956).
Madelin, M. F.: Visible changes in the vegetative mycelium of *Coprinus lagopus* Fr. at the time of fruiting. Trans. Br. Mycol. Soc. *43,* 105–110 (1960).
Marchant, R.: Wall composition of monokaryons and dikaryons of *Coprinus cinereus.* J. Gen. Microbiol. *106,* 195–199 (1978).
Matthews, T. R., Niederpruem, D. J.: Differentiation in *Coprinus lagopus.* II. Histology and ultrastructural aspects of developing primordia. Arch. Mikrobiol. *88,* 169–180 (1973).
McLaughlin, D. J.: Ultrastructural localization of carbohydrate in the hymenium and subhymenium of *Coprinus.* Evidence for the function of the Golgi apparatus. Protoplasma *82,* 341–364 (1974).
Michalenko, G. O., Hohl, H. R., Rast, D.: Chemistry and architecture of the mycelial wall of *Agaricus bisporus.* J. Gen. Microbiol. *92,* 251–262 (1976).
Moore, D.: Production of *Coprinus* protoplasts by use of chitinase or helicase. Trans. Br. Mycol. Soc. *65,* 134–136 (1975).
Moore, D., Elhiti, M. M. Y., Butler, R. D.: Morphogenesis of the carpophore of *Coprinus cinereus.* New Phytol. *83,* 695–722 (1979).
Rao, P. S., Niederpruem, D. J.: Carbohydrate metabolism during morphogenesis of *Coprinus lagopus (sensu* Buller). J. Bacteriol. *100,* 1222–1228 (1969).
Robert, J. C.: Fruiting in *Coprinus congregatus:* biochemical changes in fruitbodies during morphogenesis. Trans. Br. Mycol. Soc. *68,* 379–387 (1977a).
Robert, J. C.: Fruiting in *Coprinus congregatus:* relationship to biochemical changes in the whole culture. Trans. Br. Mycol. Soc. *68,* 389–395 (1977b).
Tracey, M. V.: Chitin. In: Modern Methods of Plant Analysis, Vol. 2. Paech, K., Tracey, M. V., (eds.). Berlin: Springer-Verlag 1955, pp. 264–274.
van der Valk, P., Marchant, R., Wessels, J. G. H.: Ultrastructural localization of polysaccharides in the wall and septum of the basidiomycete *Schizophyllum commune.* Exp. Mycol. *1,* 69–82 (1977).
Wong, W. M., Gruen, H. E.: Changes in cell size and nuclear number during elongation of *Flammulina velutipes* fruit bodies. Mycologia *69,* 899–913 (1977).

Author Index

Ainsworth, G. C.   1, 2, 6
Aist, J. R.   49, 71
Aldrich, H. C.   48, 74
Allen, J. V.   42, 43, 69, 70
Almeida, M. G.   126, 155
Almoslechner, E.   135, 152
Andrews, J. H.   1, 6
Aschan-Åberg, K.   136, 152
Ashton, M. L.   51, 52, 69

Baldy, P.   137, 154–155
Bandoni, R. J.   15, 24, 34
Banno, J.   24, 34
Barksdale, A. W.   125, 152
Bartnicki-Garcia, S.   55, 59, 69
Beckett, A.   38, 40, 51, 56, 69, 73, 166, 172
Beever, R. E.   46, 69
Bell, W. R.   97, 111
Bennell, A. R.   42, 43, 56, 69
Berger, H.   100, 111
Bernstein, H.   100, 111
Billar-Palasi, C.   118, 122
Bistis, G.   125, 152, 153
Blayney, G. P.   168, 170, 171, 172
Blumenthal, H. J.   46, 71
Bonner, J. T.   61, 62, 69, 170, 172
Borriss, H.   5, 6, 126, 130, 131, 133, 149, 153
Bouillenne-Walrand, M.   135, 153
Bracker, C. E.   38, 51, 59, 69, 70
Brefeld, O.   1, 6, 25, 34, 133, 153
Bret, J. P.   132–134, 149, 153, 168, 172
Brezden, S. A.   24, 34
Broker, T. R.   101, 111
Bronchart, R.   46, 49, 69
Brower, D. L.   62, 69

Buller, A. H. R.   61, 69, 127, 153
Bu'Lock, J. D.   125, 137, 153
Burge, H. A.   51, 52, 69
Burnett, J. H.   171, 173
Bushnell, J. L.   46, 71
Butcher, R. W.   116, 122
Butler, G. M.   125, 153
Butler, R. D.   157, 173

Calonge, F. D.   49, 69
Campbell, R.   51, 69
Catcheside, D. G.   99, 111
Caten, C. E.   89, 91
Chiquoine, A. D.   170, 172
Chiu, S. M.   95, 100, 108, 109, 112
Clark, A. J.   101, 111
Clemençon, H.   49, 52, 53, 56, 59, 69, 70
Coffey, M. D.   38, 70
Cole, G. T.   59, 73
Corner, E. J. H.   2, 6, 17, 28, 34, 37, 38, 40, 48, 49, 53–55, 61, 62, 68, 70
Couch, J. N.   12, 34
Cox, R. J.   132, 153, 168, 172
Craig, G. D.   53, 54, 70, 133, 153, 168, 172
Crawford, D. A.   28, 34
Cutter, V. M., Jr.   24, 35

Dangeard, P. A.   3, 6
Day, A. W.   89, 90
de Bary, A.   2, 6
Demoulin, V.   46, 49, 69
Dempsey, G. P.   46, 69
Derx, H. G.   25, 34
Dodge, B. O.   17, 35
Donk, M. A.   12, 34, 43, 46, 70
Doquet, G.   24, 34
Durand, R. J. C.   133, 153

Eilers, F. I.   132, 133, 149, 150, 153, 168, 172
Eisenstark, A.   101, 111
Elhiti, M. M. Y.   157, 173
Emerson, S.   99, 111
Engels, L.   135, 153
Esser, K.   61, 70, 75, 90, 113, 122
Eudall, R.   43, 71

Farr, A. L.   158, 173
Feder, N.   41, 70
Fell, J. W.   24, 34
Fields, W. G.   54, 70
Fischer, D. B.   41, 70
Flegler, S. L.   54, 70
Fogel, S.   99, 111
Friend, D. S.   59, 70
Frost, L. C.   100, 112
Fuller, M. S.   85, 89, 90
Fulton, C.   89, 90

Garcia Mendoza, C.   51, 70
Gardner, J. S.   42, 43, 70
Gäumann, E.   15, 34
Giddings, T. H.   62, 69
Gilman, A. G.   116, 122
Girbardt, M.   55, 70, 89, 90
Gooday, G. W.   131, 132, 134, 149, 153, 157, 158, 161, 166–168, 172, 173
Gould, C. J.   15, 34
Gräntz, F.   126, 133, 153
Grell, R. F.   100, 111
Greulach, V. A.   127, 136, 154
Greuter, B.   49, 70
Grove, S. N.   38, 70
Gruen, H. E.   127–130, 133–140, 142, 143, 151–155, 168, 170, 173
Guinta, C.   59, 73
Gull, K.   53–55, 61, 69–71, 75, 76, 89, 90, 133, 153, 168, 172
Gyurkó, P.   133, 154

Hädrich, H.   55, 70, 89, 90
Hagemann, F.   133, 155
Hagimoto, H.   127, 128, 135, 136, 150, 154
Hamilton, I. R.   116, 122
Hamlett, N. V.   100, 111
Hammond, J. B. W.   170, 173
Harder, D. E.   38, 42, 43, 46, 71, 89, 90

Harper, R. A.   4, 6
Hashimoto, T.   46, 71
Hastings, P. J.   99, 111
Heath, I. B.   38, 40, 51, 56, 61, 69, 71, 166, 172
Heath, M. C.   38, 42, 43, 61, 71, 72
Heintz, C. E.   66, 71, 76, 91
Henderson, D. M.   42, 43, 56, 69, 71
Henderson, S. A.   100, 111
Hepler, P. K.   55, 64, 71
Hess, W. M.   38, 39, 42, 43, 46, 69–72
Hoch, H. C.   48, 49, 52, 54, 56, 59, 71, 73, 75, 76, 79, 89, 91
Hoffman, A. A.   170, 172
Hohl, H. R.   171, 173
Hollenstein, G. O.   51, 73
Holliday, R.   99, 105, 107, 111
Hooper, G. R.   54, 70
Howard, R. J.   49, 71
Hugueney, R.   48, 52, 53, 71
Hunsley, D.   167, 168, 171, 173
Hunter, J. L.   24, 34

Ishikawa, T.   113–123, 136, 155
Ishizaki, H.   43, 71

Jaffe, L. S.   62, 64, 71
Jay, R. F.   89, 90
Jeffreys, D. B.   127, 136, 154
Jeng, D. Y.   93, 97, 112
Jensen, W. A.   41, 71
Jersild, R.   76, 91
Jirjis, R. I.   170, 173
Johri, B. N.   15, 34

Kamada, T.   133, 135, 154, 155, 171, 173
Keeble, S. A.   89, 90
Keller, J.   49, 53, 71
Kennedy, L. L.   56, 74
Khan, S. R.   46, 53, 54, 71
Khandelwal, R. L.   116, 122
Kimbrough, J. W.   48, 54, 71, 74
Kipnis, D. M.   116, 122
Kitamoto, Y.   134, 154, 170, 173
Knoll, F.   126, 127, 154
Kohno, M.   42, 43, 71
Kollmorgen, J. F.   39, 43, 72
Konishi, M.   127, 128, 135, 136, 150, 154
Kota, M.   136, 155
Kozova, J.   148, 154

Krebs, E. G.   117, 123
Krisch, H. M.   100, 111
Kubai, D. F.   89, 90
Kuenen, R.   75, 90
Kühner, R.   49, 51, 53, 72
Kunoh, H.   42, 43, 71
Kushner, S.   100−112
Kwon-Chung, K. J.   24, 35

Laffin, R. J.   24, 35
Lamb, B. C.   100, 112
Landner, L.   100, 112
Larner, J.   118, 122
Larpent, J. P.   125, 154
Latché, J. C.   137, 154
Leal, J. A.   51, 70
Lerbs, V.   56, 72, 85, 90
Léveillé, J. H.   1, 6
Littlefield, L. J.   38, 42, 43, 72
Lobo, K. J.   24, 34
Lowry, D. H.   158, 173
Lowy, B.   2, 6, 21, 35
Lu, B. C.   3, 6, 75, 76, 87, 89, 91, 93−95, 97, 100, 103, 105−109, 112
Lyons, J. M.   64, 72

Madelin, M. F.   67, 72, 168, 170, 173
Magnus, W.   126, 154
Maguire, M. P.   100, 112
Malençon, G.   21, 35
Manachère, G.   133, 154
Manocha, M. S.   61, 72
Marchant, R.   79, 91, 168, 170−173
Martin, G. W.   2, 6, 18, 35
Matthews, T. R.   137, 154, 170, 173
McLaughlin, E. G.   54, 72
McLaughlin, D. J.   3, 6, 37, 38, 40, 42, 45, 46, 48−61, 64−67, 69, 72, 75, 89, 91, 166, 170, 172, 173
McNabb, R. F. R.   2, 7
McNelly-Ingle, C. A.   100, 112
Michalenko, G. O.   129, 136, 154, 171, 173
Mims, C. W.   38, 42, 43, 46, 72
Miyazaki, S.   135, 154, 171, 173
Moens, P. B.   51, 52, 69
Mollenhauer, H. H.   40, 56, 59, 72
Möller, A.   18, 35
Moore, D.   157, 170, 171, 173
Moore, R. T.   79, 87, 91

Moran, D. T.   65, 73
Morioka, W. T.   170, 172
Morré, D. J.   40, 55, 59, 72
Mortimer, R. K.   99, 111
Müller, L. Y.   42, 72
Murray, M. J.   59, 70

Nakai, Y.   44, 52, 54, 60, 61, 72
Naudy-de Serres, M.   137, 154
Newell, S. Y.   24, 34
Newsam, R. J.   53, 54, 61, 69, 71, 75, 76, 89, 90
Nichols, R.   170, 173
Niederpruem, D. J.   66, 71, 76, 91, 132, 137, 153, 154, 168, 170, 172, 173
Nielsen, N.   135, 155
Nishimura, T.   42, 43, 71
Novaes-Ledieu, M.   51, 70
Nuccitelli, R.   62, 64, 71
Nyland, G.   24, 35

Oberwinkler, F.   21, 28, 30, 35
O'Brien, T. P.   41, 70
O'Donnell, K.   38, 55
Oláh, G. M.   40, 52, 53, 59, 73
Olive, L. S.   3, 7, 21, 24, 25, 35, 89, 91

Palevitz, B. A.   55, 64, 71
Panzica, G.   59, 73
Panzica-Viglietti, C.   59, 73
Parker, C. W.   116, 122
Patouillard, N.   1, 2, 7
Patton, R. F.   56, 73, 75, 76, 79, 89, 91
Peacock, W. J.   100, 112
Pegg, G. F.   136, 155
Pegler, D. N.   52, 53
Perkins, J. P.   117, 123
Perreau, J.   48, 49, 51−53, 73
Perreau-Bertrand, J.   49, 73
Phaff, H. J.   24, 34, 35
Pickett-Heaps, J. D.   89, 91
Pinto-Lopes, J.   126, 155
Plough, H. H.   100, 112
Plunkett, B. E.   137, 155
Pozsár, B. I.   136, 155
Prentice, H. T.   42, 43, 56, 69, 71

Radding, C. M.   101, 112
Raju, N. B.   89, 91, 93, 100, 112

Ramberg, J. E.   38, 40, 42, 43, 46, 48, 54–56, 73
Randall, R. J.   158, 173
Rao, P. S.   170, 173
Raper, C. A.   113, 122
Raper, J. R.   113, 122
Rast, D.   49, 51, 70, 73, 171, 173
Řeháček, Z.   148, 154
Reijnders, A. F.   133, 155
Reisinger, O.   52, 53, 59, 73
Rijkenberg, F. H. J.   42, 72
Robb, J.   38–40, 42, 46, 73
Robert, J. C.   134, 155, 168, 171–173
Robinow, C. F.   89, 91
Robinson, K. R.   62, 71
Rosebrough, N. J.   158, 173
Rosen, F.   3, 7
Rosen, O. M.   117, 118, 122
Rosenvinge, L. K.   3, 7
Rossen, J. M.   97, 112
Rowley, J. C.   65, 73
Rubin, C. S.   117, 118, 122
Ruiz-Herrera, J.   59, 69
Rypáček, V.   136, 155

Sapin-Trouffy, P.   3, 6
Savile, D. B. O.   51, 73
Scannerini, S.   59, 73
Schmitz, J.   5, 7, 126, 149, 155
Schroeter, J.   17, 35
Schwantes, H. O.   133, 155
Seabury, F.   38, 42, 43, 46, 72
Semerdžieva, M.   61, 70
Setliff, E. C.   48, 49, 52, 54, 56, 59–61, 71, 73, 75, 76, 79, 89, 91
Shear, C. L.   17, 35
Simchen, G.   100, 112
Sladký, Z.   136, 155
Spurr, A. R.   76, 91
Stadler, D. R.   99, 100, 112
Stahl, U.   61, 70, 113, 122
Stamberg, J.   100, 112
Statzell, A. C.   24, 34
Steiner, A. L.   116, 122
Streeter, S. G.   126, 155
Sundberg, W. J.   38, 40, 49, 60, 61, 73
Sutherland, E. W.   116, 122
Syrop, M. J.   51, 73
Szabó, L. G.   136, 155

Takemaru, T.   133, 135, 154, 155, 171, 173
Talbot, P. H. B.   1, 2, 7, 46, 54, 71
Therrien, C. D.   97, 111
Thielke, C.   3, 7, 53, 54, 56, 60, 61, 74–76, 83, 87, 91
Thurston, E. L.   38, 42, 43, 46, 72
Tichý, V.   136, 155
Towe, A. M.   100, 112
Tracey, M. V.   158, 166, 173
Trione, E. J.   39, 43, 72
Truter, S. J.   42, 72
Tsuneda, I.   56, 74
Tu, C. C.   48, 74
Tulasne, C.   1, 7
Tulasne, E. L.   1, 7
Turner, E. M.   135, 155

Uno, I.   113–123, 136, 155
Urayama, T.   127, 135, 136, 155
Ushiyama, R.   44, 52, 54, 60, 61, 72

van den Ende, H.   125, 155
van der Valk, P.   59, 74, 171, 173
van der Walt, J. P.   24, 35
Volz, P. A.   76, 91

Wager, H.   3, 7
Wakayama, K.   3, 7
Walsh, D. A.   117, 123
Weber, D. J.   38, 39, 42, 43, 46, 69, 71
Weese, J.   17, 35
Wells, K.   3, 4, 7, 38, 40, 46, 48, 52–54, 56, 61, 74, 75, 77, 85, 89, 91
Wessels, J. G. H.   59, 74, 171, 173
Westergaard, M.   97, 112
Whitehouse, H. L. K.   99, 112
Willam, A.   135, 153
Wong, W. M.   133, 134, 137, 155, 168, 173
Wood, D. A.   133, 153, 168, 172
Wu, S.   134, 136, 142, 151, 153, 154
Wu-Yuan, C. D.   46, 71

Yamaguchi, M.   116, 117, 123
Yoon, K. S.   48, 52, 53, 67, 74
Young, T. W. K.   52, 53, 73

Zieg, J.   100, 112

Subject Index

N-Acetylglucosamine   164, 166, 167, 171
UDP-N-Acetylglucosamine   166
Achlya   125
Adenosine 3'-monophosphate
   (3'-AMP)   *see* Basidiocarp,
   initiation; Stipe, elongation
Adenosine 3',5'-cyclic monophosphate
   (cyclic AMP; cAMP)   *see*
   Basidiocarp, initiation; Stipe,
   elongation
   dependent protein kinase   117–119,
      121, 122
Adenosine 5'-monophosphate
   (5'-AMP)   115, 116, 136; *see also*
   Stipe, elongation
Adenylate cyclase   115–117, 120
Aeciospore   43
Aessosporon   9, 24
Agaricaceae   33
Agaricales   1, 5, 10, 28–30, 32, 33, 37,
   52, 56, 75, 76, 85, 93, 136, 149
Agaricus   10, 51, 126, 127
A. bisporus   5, 51, 53, 56, 59, 75, 76, 83,
   126–129, 132–136, 139, 149, 150,
   152, 168, 170
A. campestris   61, 126
A. xanthoderma   32
Albatrellus   10, 30
Amanitaceae   33
Amanita crenulata   126
3'-AMP   *see* Adenosine
   3'-monophosphate
5'-AMP   *see* Adenosine
   5'-monophosphate
Aphelaria   10, 26
Aphelaria-like   28

Aphyllophorales   5, 10, 21, 26, 28, 30,
   37
Arabitol   134
Armillaria matsutake   *see Tricholoma
   matsutake*
Artifact, fixation   4, 77, 80, 89
Ascobolus stercorarius   125
Ascomycetes   51, 85
Ascomycotina   4, 15
Ascospore-delimiting membrane   51
Ascospore wall   51
Ascus vesicle   51
Athelia   10, 27, 30
A. epiphylla   30
Auricularia   42
A. fuscosuccinea   2, 3, 42, 46–48, 54,
   55, 61, 67, 68
Auriculariales   9, 11–13, 15, 17–20,
   24, 25, 28, 30
Auxin   135, 136
Avena coleoptile   1

Basidiocarp   5
   agaricoid   18
   clavarioid   15, 17, 21, 30
   corticioid   17, 18, 21, 30
   development   *see also* Stipe, curvature;
      Stipe, elongation
      metabolic control   157–173
   dry weight   *see also* Stipe, elongation
      increase   134, 135
      loss   134
   evolution   21, 30
   excised   *see* Stipe
   fossil   9
   gastroid   17, 18, 24

Basidiocarp *(cont.)*
  hydnoid 18
  initiation 4, 5, 113–123
    catabolite repression 117, 121
    effect of
      adenosine 3'-monophosphate 113–116, 136
      adenosine 3',5'-cyclic monophosphate 4, 113–123, 136, 137
      basidiocarp extract 113–115, 117, 120, 121
    monokaryotic 61, 113–123
  morphogenesis *see* Basidiocarp, development
  odontoid 18
  pustulate 21
Basidiomycotina 1, 3, 5, 9, 13
Basidiospore 9, 37–74
  apiculus *see* Basidiospore, hilar appendix
  apophysis 52
  asymmetry 49–51
  cytochemistry 37–74
  cytoplasm 50, 52
  development 2, 37–74
  discharge 40, 52, 53
    active 12, 15, 20
    apparatus 52
    passive 10, 12, 15, 20, 21; *see also* Basidium, gastroid
  gasteromycete-like 12, 53
  germination
    by repetition *see* Secondary spore
    phylogenetic significance 1, 2
    yeast-like 23
  hilar appendage 53
  hilar appendix 50, 52, 53
    droplet 52, 53
  hilar appendix body 50, 52
  hilar plug 52, 53
  hilum 52, 53
  initiation 38, 41, 50
  morphology 49, 51
  ontogeny *see* Basidiospore, development
  phylogeny 52
  primordium *see* Basidiospore, apophysis
  punctum lacrymans 53

  ultrastructure 37–74
  wall 3, 49, 51, 52
    exospore, Agaricales 52
    glucan layer 51, 59
    phylogenetic interest 51
    setting 49
    terminology 49
Basidium 9–35, 37–74
  ampoule effect 54, 55, 68
  apically expanding 30, 32
  "charging" 38, 40, 68
  clavate 30
  cytochemistry 37–74
  cytoplasm 37, 38, 40, 55, 67
    movement 55
    organization 38, 49
  development 2, 3, 18, 37–74
    experimental control 61–64
    methods 66, 67
    number/pattern of sterigmata 61
      effect of electrical field 62, 66, 68
      effect of number and/or position of nuclei 61
      effect of temperature 62–64, 68
    sterigmal initiation 38, 61, 62
      effect of electrical field 62, 65, 68
      effect of temperature 62, 68
    stages 41
    synchrony in *Coprinus* *see* Meiosis, synchrony in *Coprinus*
  dimorphic 63, 76
  epibasidium 48
  gastroid 10–13, 15–19, 21, 24, 30, 33; *see also* Basidiospore, discharge, passive
  heterobasidium 2; *see also* Basidium, phragmobasidium
  holobasidium 10, 20, 21, 26, 28, 30, 31, 37, 38, 42, 46, 48, 54, 55, 61, 62, 67, 68
  homobasidium 2
  hymenomycetoid 12, 13; *see also* Basidiospore, discharge, active
  initiation 37–40
  membrane *see* Endomembrane system
  meruliaceous 30, 31

Subject Index

metabasidium 3, 16, 17, 38−40, 42, 43, 46, 48, 56
morphogenesis *see* Basidium, development
morphology, comparative 1, 2, 26
non-inflating 30, 32
ontogeny *see* Basidium, development
phragmobasidium 10, 20, 26, 28, 37, 38, 42, 46, 47, 54, 55, 67, 68
phylogenetic significance 2, 3, 9−35, 37
pleurobasidium 26−29
postmeiotic 38, 57, 60, 78, 79, 85
prefusion 38, 40, 41
premeiotic 70, 77
probasidium 2, 3, 12, 13, 15, 17, 21, 25, 38; *see also* Teliospore wall 42
promycelium *see* Basidium, metabasidium
protosterigma *see* Basidium, epibasidium
septate *see* Basidium, phragmobasidium
septation 46−48
septum 3, 47, 68
  adventitious *see* Basidium, septum, secondary
  basal 40, 46, 47, 53
  partial 20−22, 26, 28
  primary 46, 48
  reduction 26
  secondary 46, 48
single-spored 25
sphaeropedunculate 18, 19
stichobasidium 61
storage products 2, 3, 40−42
suburniform 26, 28
terminology 2, 12
turgor pressure 55
type 10, 11, 13−15, 20−23, 26, 28, 32, 33
ultrastructure 37−74
urniform 26, 28, 30, 31
urnigera-type *see* Basidium, urniform
vacuolation 38, 48, 54, 55
  related to ecological adaptations 55
wall 37, 42, 68
  development 3, 42−46
6-Benzylaminopurine 145, 146

Bolbitiaceae 33
Boletaceae 33
Boletales 10
Boletus edulis 135
B. rubinellus 40, 44, 46, 48, 52, 53, 56, 67, 75
Botryobasidium 10, 26−28
Brachybasidium 10, 21
B. pinangae 23

Caffeine 115, 116
Calocera 9, 21
Candida albicans 148
Cantharellales 10
Carbohydrates 40, 41, 59
  changes in basidiocarps 135
  transport into basidiocarps 134, 135
Catabolite repression *see* Basidiocarp, initiation
Cellular program 108, 109
Centriole equivalent *see* Spindle pole body
Centriole-like organelle *see* Spindle pole body
Ceratobasidiaceae 28
Ceratobasidium 9, 21
C. calosporum 22
Cerinomyces 9, 21
C. crustulinus 22
Chitin 51; *see also* Stipe, elongation
Chitin synthetase 59
Chromatin *see* Chromosome
Chromosome 79, 80, 82−84, 86, 89, 90, 106; *see also* Meiosis; Mitosis
  affinity for nuclear envelope 4, 80, 89
  asynchronous disjunction 4, 83
  morphology during division 3, 75, 89
  ring structure 79, 89
Chrysomyxa 9
C. abietis 12
Chrysomyxaceae 12
Clamp connection, loop-like, sub-basidial 30, 31
Clavaria 10, 30
C. tenuipes 30
Coleosporiaceae 12
Coleosporium 12, 13
C. sonchi 16
Coniodictyum 10
C. chevalieri 23

Coprinaceae 33
Coprinus 4, 5, 10, 37, 52, 75–77, 93, 126, 127, 131, 134, 135, 150, 170, 172
Coprinus spp. 52
C. cinereus 2, 4, 37–68, 75, 78–80, 83–85, 87, 93–111, 131, 132, 134, 149, 157, 159, 161, 165–168, 170, 171
C. comatus 32
C. congregatus 132–134, 149, 168, 171, 172
C. ephemerus 133
C. lagopus see C. cinereus; C. radiatus
C. macrorhizus 4, 113–121, 133, 135, 136
C. macrorhizus f. microsporus 113, 127, 136, 157, 171
C. micaceus 75, 82, 86–88, 136
C. radiatus 56, 75, 83, 85, 126, 132, 133, 137, 149, 150, 168
C. stercorarius 133
C. sterquilinus 126, 127, 136
Corticiaceae 28, 29
Cortinariaceae 33
Crossing over 93–112
  potential 106, 107
Cryptobasidiales 10–12, 20, 21, 23, 24
Cryptococcales 9, 24
Cyclic AMP see Adenosine 3′, 5′-cyclic monophosphate
Cyphellostereum 10, 30
Cystidium, wall 42
Cytokinin 136, 152
Cytokinin-like activity 136

Dacrymyces 9, 21
D. ovisporus 25
D. punctiformis 22
D. stillatus 54
Dacrymycetales 9–11, 20, 21, 24, 25, 30
Deoxyribonucleic acid (DNA)
  chain elongation 93, 94, 98
  model, hybrid 99
  nicking 4, 101, 102, 106, 107
    coincidental 106, 107
    effect of temperature 102
    enzyme 108
    program 108, 109, 111

^{32}P-labeling 4, 97–99, 101, 102, 104
  repair 4, 100–106
    program 108, 109, 111
  replication 4, 93–112
    restrictive conditions 4, 93–98
  replicons 98
  synthesis 93, 97, 99, 101; see also Deoxyribonucleic acid, replication
Dicellomyces 10, 21
Digitatispora 9
D. marina 23, 24
"Digitatisporales" 9
Dolipore see Septal pore apparatus

Endomembrane system 55–60, 66
Endoplasmic reticulum (ER) 38–40, 42, 54–57, 59, 60, 85
  perinuclear cisternae 56
  rough 50, 56
    secretory function 56
  vesicles 55, 59
Entoloma saundersii 75
Eocronartium 9, 12, 13, 15, 17
E. muscicola 15
Euascomycetes 51, 52
Exidiopsis 5
Exobasidiales 10–12, 20–24
Exobasidiellum 10, 21
E. graminicola 23
Exobasidium 10, 21
E. oxycocci 23

Filobasidiella 9, 24
Filobasidium 9, 24
Flammulina velutipes 5, 75, 76, 80, 126, 129–134, 136–141, 143–152, 168, 170
Fomes fomentarius 56
Fruitbody see Basidiocarp

Ganoderma 5
Gasteromycetes 28–30, 49, 53
  secotiaceous 33
Gastrosporiales 10
Gautieriales 10, 30, 33
Geastrales 10, 30, 33
Gene conversion 99
Gibberellic acid 136; see also Stipe, elongation
Gibberellin 136

Gibberellin-like activity 136
Gill *see* Lamella
β-Glucan 51
Glycogen 3, 39, 40, 60
  degradation 134
Glycogen phosphorylase 118–121
Glycogen synthetase 118–120
Golgi apparatus 55–57
  cisternae 49, 50, 56–60, 68
    plate 57–59, 68
    ring 58, 59, 68
    tubular-vesicular 58, 59, 68
  dictyosomes 56, 59, 87
  staining 59
  vesicles 57–60
Gomphidiaceae 33
Goplana 9, 12, 13
G. micheliae 16
Graphiolales 9
Gymnosporangium
    juniperi-virginianae 43

Hemiascomycetes 52
Herpobasidium 9, 13, 17
H. deformans 15
H. filicinum 15
Heterobasidiomycetes 1–3, 9, 10, 12, 21, 22, 24, 25
Heteroduplex formation 105, 107
Hoehnelomyces 9, 17
Holobasidiate taxa 21
Holobasidiomycetes 21, 51
Homobasidiomycetes 1–3, 9, 10, 26, 28, 30, 75
Humidicutis 10, 30
H. marginatus 30
Hyaloria 9, 18
H. pilacre 19
Hydnopolyporus 10, 30
Hymenium 40, 42, 44, 45, 63, 65
  hyphal wall 42
  pellicle 39, 43–46, 48, 67
  rodlet surface pattern 44–48, 67
Hymenochaetales 10
Hymenogastrales 10, 33
Hymenomycetes 49, 53, 59
Hyphal tip organization 38, 49
Hyphoderma 10, 28
Hyphodontia 10, 26–28, 30

Indol-3-acetic acid (IAA) 135, 136; *see also* Stipe, elongation
Itersonilia 9, 24
I. perplexans 25

Jola 15

Karyogamy 3, 4, 37, 56, 76, 94–106, 108–110
Kinetin 136
Kinetochore 76, 79, 83, 90

Laccaria amethystina 75
Lamella (*see also* Hymenium)
  diffusate, effect on stipe elongation; *see also* Pileus expansion; Stipe, curvature; Stipe, elongation
  absence in pileus trama 128, 142
  decrease after prolonged diffusion 148, 151
  decrease in aging basidiocarps 130, 140, 141
  detected in *Agaricus bisporus* 127, 129, 150
  diffusible 145–147
  heat stable 146–148, 152
  increase by potato extract 130, 142, 143, 151
  persistence 148
  reduced when tested in agar rings on lateral stipe surface 143, 144, 151
  intercellular matrix 43, 44
  pellicle *see* Hymenium, pellicle
  trama 40
    hyphal wall 42
Lentinus 10
L. elodes 44, 52, 61, 75–77, 81, 114, 115
L. tigrinus 32, 133, 135, 136
Lepista nuda 75
Leucosporidium 9, 24
Light green, diffusion into stipes 144, 145
Lipids 3, 40, 41
  droplets 38, 40, 42, 47, 50, 60
Lomasome 61, 79, 80, 166, 168, 169
Lycoperdales 10, 33
Lysosome 55, 59

Mannitol 134
Meiosis 2–4, 37, 38, 56, 75–91; see
    also Chromosomes;
    Deoxyribonucleic acid;
    Recombination; Spindle pole body
  anaphase 4, 82, 83
  asynchronous 76, 93
  cell cycle 93, 95, 108, 109
  diplotene 94–96, 102, 103, 109
  metaphase I 79, 83, 84, 104, 105
  pachytene 4, 76, 94–96, 100–106,
    108–110
  program 108, 109
  prophase 3
  reduction in volume 56
  S-phase 4, 93–100, 102, 108, 109
  synchrony in *Coprinus* 4, 75, 93, 101,
    111
  telophase 83
Meiosporangium *see* Basidium
Meiospore *see* Basidiospore
Melampsorella 13
Melanin 51
Melanogastrales 10, 30, 33
Merulius 10, 30
Method
  culture of basidiocarps
    Auricularia fuscosuccinia 67
    Boletus rubinellus 67
    Coprinus cinereus 64, 94
    Flammulina velutipes 137, 138
  cytochemical 64, 66
  detection
    carbohydrates 58, 59, 66
    chitin 158
    endomembrane system 66
    glucose 157, 158
    glycogen 158
    protein 42, 64, 158
    ribonucleic acid 64
    trehalose 157, 158
  electron microscopy 4, 49, 64, 76
  freeze-etch 66
  high-voltage electron microscopy 65
  iron haematoxylin nuclear stain 94
  selection of test stipes in *Flammulina
    velutipes* 137–139
  vital staining 89
    coriphosphin 89
Microbodies 50, 59, 61

Microfilaments 54, 55
Microstroma 10
M. juglandis 23
Microtubules 49, 60, 61, 64, 79, 82–87,
  89
Microtubule organizing center 89; *see
  also* Spindle pole body
Mitochondrion 38, 39, 61
Mitosis 61, 75
  postmeiotic 85
  two-track model 89
Mucidula 10
M. mucida 32
Mucopolysaccharides 59
Multiclavula 10, 30
M. mucida 30
Multivesicular bodies 59, 61
Myxarium 9, 18, 19
M. nucleatum 46, 48, 56

Nematoloma fasciculare 127
Neottiella rutilans 97
Neurospora crassa 46
Nidulariales 10, 30, 33
Nitrogen, organic, increase in
  basidiocarps 134
Nuclear cycle 109
Nucleolus 76, 79, 81, 89
Nucleus 39, 57, 60, 89
  envelope 3, 55, 56, 59, 60, 75, 76,
    79–90
  polar gaps 85
  migrating 55, 60, 61, 77,
    85
  post-meiotic 56, 88
Nucleus-associated organelle *see* Spindle
  pole body

Ochropsora 9, 13
O. sorbi 12, 16
Oliveonia 9, 26, 28
Omphalina 10, 30
O. lutealilacina 30

Panaeolina foenisecci 75
Panellus 10, 30
P. violaceofulvus 30
Panus tigrinus see Lentinus tigrinus
Paraphelaria 9
P. amboinensis 17

Paraphysis 44, 46, 78, 79
    wall 42
Patouillardina 9, 18, 19
P. cinerea 19
Paullicorticium 10, 26–28
Paxillaceae 33
Pellicle *see* Hymenium, pellicle
Phallales 10, 30, 33, 53
Phanerochaete 10, 30
Phlebia 10, 27, 30
P. radiata 30
Phleogena 9, 17
Pholiota terrestris 54
Phosphodiesterase 115–117, 120, 121
Pilacrella 9, 13, 17
P. solani 15, 17
Pileus expansion, dependent on
    lamellae 127, 128, 150
Plasmalemmasomes 79
Plasma membrane 51, 52, 55, 56, 64
Platygloea 9
P. unispora 25
Pleurotus ostreatus 75, 133
Polyporaceae 33
Polyporales 10
Polysaccharides 42, 53, 59
    degraded in basidiocarps 134
Poria latemarginata 56, 75, 79
Porodisculus 10, 30
Potato extract *see* Lamella, diffusate
Protein 40, 41, 51, 54, 67
    kinase 118
    secretion 56
Protobasidiomycete 51
Protodontia 9, 18
P. uda 19
Protohymeniales 10
Psathyrella disseminata 127
Pseudohydnum 9, 18
Pseudoparaphysis *see* Paraphysis
Pseudotulasnella 9, 21, 28
P. guatemalensis 22
Psilocybe turficola 75, 81
Pterula sp. 54
Puccinia 5, 9, 89
P. caeomatiformis 12, 16
P. coronata f. sp. *avenae* 42, 43
P. malvacearum 38, 55
P. podophylli 43

P. smyrnii 42, 43, 56
Pucciniastraceae 12

Receptive hyphae 12
Recombination 93, 96, 99–101,
    104–107, 109, 111; *see also*
    Crossing over
    coordinated program, hypothesis 4,
        108, 109
    function of time 105
    mode, hypothesis 105–107
    temperature effect 4, 100–110
Repetobasidium 10, 26–28
Replication *see* Deoxyribonucleic acid
Rhizopus sp. 158
Rhodophyllaceae 33
Rhodosporidium 9, 24
Ribonucleic acid (RNA) 40, 41, 54, 67
Ribosomes 38–40, 49, 54, 56, 58, 79,
    85, 88
Russula 10
Russula spp. 52
R. mairei 32
Russulaceae 33
Russulales 10
Rusts *see* Uredinales

Schizophyllum commune 40, 52, 59
Sclerodermatales 10, 30, 33
Sebacina 26
Sebacina sp. 22
Secondary spore 9, 10, 26, 28
Septal pore apparatus 3, 12, 46–48, 53,
    54, 68
    cap 53, 54
    hymenial/subhymenial 53
    outer cap 3, 47, 53, 54, 68
        function 54
        microfilaments 54
    pore 54
        membrane 54
        occlusions 47, 54
        rings 54
        sealing 54
    septum 47
    swelling 47, 54
Septobasidiales 9, 11–13, 17, 21, 24
Septobasidium 9, 13
S. albidum 17
Siderophilous granules 59

Sistotrema 10, 26−28, 30
S. brinkmanni 30
Smuts *see* Ustilaginales
Smut spores 15, 23
Sordaria fimicola 97
Spermogonium 12
Spindle apparatus 3, 75, 79, 85, 89, 90
Spindle pole body 3, 4, 52, 55, 60, 61, 76, 78−86, 88−90
 function 89
 terminology 89
Spore *see* Basidiospore
Sporidiobolus 9, 24
Sporobolomycetales 9
Sporogenesis *see* Basidiospore, development
Stereophyllum 10, 30
Sterigma 20, 21, 26, 37, 44, 46, 50, 52
 appendage 53
 bifurcate 26, 28
 cytoplasm 49
 development 45, 55
  experimental control *see* Basidium, development, experimental control
 initiation 38, 45, 48, 49
 phylogeny 20, 21, 26
 plug 52
 reduced 20
 single 25
 supernumerous 26, 28
 wall 48, 51, 68
Stilbum 9, 17
Stipe
 curvature 5
  effect of
   asymmetrical pileus remnants 5, 126−129
   lamellar diffusate 127−129
   lamellar extract 127
   lanolin 129, 130
 elongation 1, 5, 125−155, 157−173; *see also* Lamella, diffusate
  chitin content 160, 163−166, 168, 171, 172
  chitin synthesis 164, 166, 168, 171
  dependence on
   lamellae 5, 128, 129
   mycelium 5, 134, 135
   pileus 5, 126−129, 132, 133, 149, 150

 dry weight 158−161, 168
 effect of
  adenosine 3'-monophosphate 145−147
  adenosine 3', 5'-cyclic monophosphate 145−147
  adenosine 5'-monophosphate 145−147
  gibberellic acid 145, 146, 152
  indoleacetic acid 136, 145−147, 152
  nucleotides 145, 146, 152
  nutrients 5, 142
  plant growth regulators 145, 146, 152
  rudimentary pilei 132, 133
  2, 3, 5-triiodobenzoic acid 136
 glucose content 159, 161, 162, 164, 170, 172
 glycogen content 159, 162, 164, 170, 172
 protein content 160, 163, 164, 171, 172
 residual after decapitation 128, 129, 139, 140
 trehalose content 159, 161, 162, 164, 170, 172
 function 172
Strophariaceae 33
Stropharia rugosoannulata 75, 78, 79
Stypella 9, 18
S. papillata 19
Subhymenium 40
 hyphal wall 42
Sympodioconidium 38
Synaptonemal complex 76, 81, 94, 103−105, 109

Taphrina 51
T. deformans 51
Teleutospore *see* Teliospore
Teliospore 2, 3, 12, 13, 15, 16, 38, 39, 42, 43, 46, 56, 68; *see also* Basidium, probasidium
 aecioid 43
 cytoplasm 39
 germination 38−40, 43
 primary cell 43
 septum formation 43
 wall 42
  development 3, 43

Subject Index

ornamentation 42, 43, 46
   primary 42
   secondary 42
Thanatephorus cucumeris 48
Thelephorales 10
Theophylline 115, 116, 136, 145, 146
Tilletia 10
Tilletia spp. 43
T. caries 23, 39, 43
T. controversa 43
Tilletiales 2, 3, 10, 11, 20, 21, 23, 24, 43
Tilletiaria 9, 13, 15, 24
Tonoplast 79, 80
Tranzschelia anemones 42, 43
Trehalose 134; *see also* Stipe, elongation
Tremella 25
Tremellaceae 19
Tremellales 5, 9, 11 12, 18−21, 24−26, 28, 30
Tremellodendron 9, 26, 28
Tremellodendropsis 9, 28
Tremiscus 9, 18
Tricholoma 10, 30
T. matsutake 113, 115, 127
Trichophyton mentagrophytes 46
Tulasnella 9, 21, 25
Tulasnella sp. 48
T. violea 22
Tulasnellales 9, 11, 12, 20, 21, 24−26, 28
Tulostomatales 10, 30, 33

Uredinales 2, 3, 7, 9−13, 15, 16, 21, 24, 38, 42, 43

Uredinella 9, 13
U. coccidiophaga 12, 17
Urediospore 38
   development 42
   wall 42, 43
Uromyces 13
Ustilaginaceae 15
Ustilaginales 2, 3, 9, 10, 11, 13, 15, 21, 24, 46
Ustilago 13
U. hordei 40, 46
U. maydis 38, 40, 43, 46, 48, 55, 56
Uthatobasidium 9, 26, 27

Vacuole 38, 39, 55, 57, 59, 77
Vesicles 55, 67
   in developing basidiospores 59
Vuilleminia 10, 27, 28
Vuilleminiaceae 26, 28, 29

Water repellency 46

Xenasma 10, 27, 28
Xenasmataceae 26, 28, 29
Xeromphalina 10
X. campanella 32

Yeast, basidiomycete 10, 11, 24
Yeast-like somatic phase 9, 10, 24
Yeast taxa 24
Ypsilonidium 9, 21
Y. sterigmaticum 22

Springer Series in Microbiology

Editor: **Mortimer P. Starr,** Department of Bacteriology, University of California, Davis, California, U.S.A.

**Thermophilic Microorganisms
and Life at High Temperatures**
T.D. Brock, University of Wisconsin, Madison
1978/xi, 465pp./195 illus./cloth
ISBN 0-387-**90309**-7

Bacterial Metabolism
G. Gottschalk, Universität Göttingen, Federal Republic of Germany
1979/xi, 281pp./161 illus./cloth
ISBN 0-387-**90308**-9

**Ascomycete Systematics:
The Luttrellian Concept**
D.R. Reynolds (Ed.), Natural History Museum, Los Angeles
1981/vii, 242pp./122 illus./cloth
ISBN 0-387-**90488**-3

**Bacterial and Bacteriophage Genetics:
An Introduction**
E.A. Birge, Arizona State University, Tempe
1981/xvi, 359pp./111 illus./cloth
ISBN 0-387-**90504**-9

General Nematology
A. Maggenti, University of California, Davis
1981/x, 372pp./135 illus./cloth
ISBN 0-387-**90588**-X

**Basidium and Basidocarp:
Evolution, Cytology, Function and Development**
K. Wells and E.K. Wells, University of California, Davis
1982/xii, 187pp./117 illus./cloth
ISBN 0-387-**90631**-2